T0377034

Manufacturing Cost Policy Deployment (MCPD) Profitability Scenarios

Systematic and Systemic Improvement of Manufacturing Costs

Manufacturing Cost Policy Deployment (MCPD) Profitability Scenarios

Systematic and Systemic Improvement of Manufacturing Costs

By

Alin Posteucă

Routledge
Taylor & Francis Group

A PRODUCTIVITY PRESS BOOK

First edition published in 2019
by Routledge/Productivity Press
711 Third Avenue New York, NY 10017, USA
2 Park Square, Milton Park, Abingdon, Oxon OX14 4RN, UK

© 2019 by Taylor & Francis Group, LLC
Routledge/Productivity Press is an imprint of Taylor & Francis Group, an Informa business

No claim to original U.S. Government works

Printed on acid-free paper

International Standard Book Number-13: 978-1-138-49873-0 (Hardback)
International Standard Book Number-13: 978-1-351-01575-2 (eBook)

Library of Congress Cataloging-in-Publication Data

Names: Posteuca, Alin, author.
Title: Manufacturing cost policy deployment (MCPD) profitability scenarios : systematic and systemic improvement of manufacturing costs / Alin Posteuca.
Description: New York, NY : Routledge, 2019. | Includes bibliographical references and index.
Identifiers: LCCN 2018026871 (print) | LCCN 2018028720 (ebook) | ISBN 9781351015752 (e-Book) | ISBN 9781138498730 (hardback : alk. paper)
Subjects: LCSH: Production management. | Industrial productivity. | Costs, Industrial.
Classification: LCC TS155 (ebook) | LCC TS155 .P573 2019 (print) | DDC 658.5--dc23
LC record available at https://lccn.loc.gov/2018026871

Visit the Taylor & Francis Web site at
http://www.taylorandfrancis.com

To Emiliana, Andreea and Ştefan

Contents

PART I Understanding of the MCPD System

**PART III   Profitable Improvement Projects
Implementation through MCPD for MCI
toward Minimizing Inputs**

List of Figures

List of Tables

Preface

Manufacturing profitability through productivity is, and will remain, an eternal challenge for manufacturing management. Although it is quite easy to understand the need for productivity and to identify a non-productivity state of a manufacturing flow, and although considerable efforts are being made to measure and improve productivity, offering the results of improved productivity at the level of profit and loss statement (P&L) (statement of operations; pro-forma and actually reported at the end of the year) is still a wish of the management teams. Often manufacturing companies have methods, techniques, and tools to tackle productivity, but real improvement, continuous targeting of improvements to profit, and financial quantification of results are often not met. In this context, annual and multiannual profit level planning requires targeted planning of successive productivity improvements based on current and future state of sales trends and necessary manufacturing capacities—market-driven trends (customers, suppliers and especially competitors)—and by the current state of the manufacturing flow.

Continuous identification of the steady growth directions of manufacturing profit, based on productivity gains, reducing manufacturing unit costs and increasing the sales volumes achieved and sold, requires the continuous development and implementation of a company-wide productivity master plan to translate the managerial vision and mission of productivity in concrete and ongoing activities and actions to improve the production flow productivity for each product family cost (PFC), based on the ongoing recovery of productivity policy deployment.

Going forward, fulfilling external (from sales) and internal (from manufacturing cost improvement—MCI) annual and multiannual manufacturing target profits through maximizing outputs and through minimizing inputs requires the development of contextual productivity scenarios and strategies to consistently transform the current production method. For example, in the context of reducing sales volumes, the reduction of manufacturing unit costs could not be achieved if a project to improve equipment effectiveness for MCI (e.g.: reduction/elimination of breakdown losses) is planned and implemented on a background of equipment already at overcapacity. The unnecessary increase in equipment productivity at that time by improving equipment effectiveness, in the context of a decrease in sales volumes for the products made with the equipment, could even lead to an increase in manufacturing unit costs, an increase due to the project costs for the equipment effectiveness improvement (reduction/elimination of equipment breakdown losses). Another example, in the opposite direction, in the context of a significant increase in sales volumes, the reduction of unit manufacturing costs could be better achieved if an improvement of the actual equipment cycle time is planned and achieved than a reduction in consumptions with packaging materials or a reduction in utility consumption. Thus the impact on the annual MCI goal of a productivity improvement project is different depending on the sales scenarios and, implicitly, those of the required capacity in that period and the forthcoming period. The same productivity improvement project may have a different impact on the annual MCI goal from one period to the next. In this context, the consistent approach to long-term profitability through productivity requires the development of sales scenarios and productivity strategies and the exact and timely allocation of all the resources needed to achieve systematic and systemic improvements.

Tangible and intangible effects of consistent and continuous application of the Manufacturing Cost Policy Deployment (MCPD) system (development and fulfillment of annual and multiannual MCI targets and means) are present at the levels of both maximizing outputs and minimizing inputs through scientific identification and structured implementation of annual and multiannual profitability improvement projects for one of the two possible hypostasis of a PFC for a certain period: (1) sales growth trend—the predominant need to increase productivity by improving effectiveness; or (2) the downward trend in sales—the overriding need to increase productivity by improving efficiency.

In order to reduce or eliminate discrepancies between productivity gains and the need to meet annual and multiannual manufacturing targets, the MCPD system approach in this book has two main directions: (1) the scientific establishment of MCI targets and means based on sales scenarios and required profitability; and (2) successful implementation of profitable productivity improvement projects (annual MCI means or annual *kaizen* and *kaikaku* for MCI).

In this context, the main audience of this book, the top management, middle management, and professional support staff who manage manufacturing areas will discover the practical approach of 15 annual and multiannual *kaizen* and *kaikaku* projects for MCI to meet the annual MCI goal and, implicitly, external and internal annual and multiannual manufacturing target profits through maximizing outputs and through minimizing inputs.

Therefore, the author's hope is that the MCPD system will provide a way to address successful implementation of profitable projects on the basis of the sales scenarios and necessary production capacity by consistently and continuously uncovering the hidden reserves of manufacturing profit directly from the manufacturing process related to each PFC through continuous and targeted productivity growth coordinated by the need for MCI.

Alin Posteucă

Author

Alin Posteucă is a management consultant in profitability, productivity and quality and the managing partner of Exegens Management Consultants (Romania). Prior to this position, he held top management positions in manufacturing and service companies.

His recent research includes the development of the Manufacturing Cost Policy Deployment (MCPD) system for multiannual target profit by identifying new manufacturing profit opportunities by exploiting manufacturing cost improvement (MCI) through continued productivity growth. The main purpose of the MCPD system is to meet the annual MCI goal. The annual MCI goal is the difference between the annual manufacturing target profit and the annual external profit expected from sales for each company in the group and for each product family cost of each company.

He received his PhD in industrial engineering from the Polytechnic University of Bucharest (Romania) and his PhD in managerial accounting from the Bucharest University of Economic Studies (Romania). He received his MBA from the Alexandru Ioan Cuza University of Iași, Romania. He is also a certified public accountant in Romania.

He has been actively involved in various industrial consulting and training projects for more than 20 years in Romania and he has published in various research journals and presented papers at numerous conferences regarding productivity, profitability and quality. He is the coauthor of *Manufacturing Cost Policy Deployment (MCPD) and Methods Design Concept (MDC): The Path to Competitiveness* (CRC Press, 2017). He is also the author of *Manufacturing Cost Policy Deployment (MCPD) Transformation: Uncovering Hidden Reserves of Profitability* (CRC Press, 2018).

List of Abbreviations

4M	Method, Machine, Man and Material
5G	*Gemba, Gembutsu, Genjitsu, Genri, Gensoku*
5S	Sort, Set, Shine, Standardize, Sustain
AM	Autonomous Maintenance
AMCIB	Annual Manufacturing Cash Improvement Budget
AMIB	Annual Manufacturing Improvement Budgets
atEPU	Annual Target External Profit per Unit
atIPU	Annual Target Internal Profit per Unit
atMC	Annual Target Manufacturing Cost
atMCU	Annual Target Manufacturing Cost per Unit
atNUS	Annual Target Number of Units to be Sold
atP	Annual Target Profit
atPU	Annual Target Profit per Unit
atSP	Annual Target Selling Price
atSPU	Annual Target Selling Price per Unit
CAMPT	Critical to Annual Manufacturing Profitability Tree
CBM	Condition-Based Maintenance
CCLW	Critical Costs of Losses and Waste
CCW	Critical Costs of Waste
CF	Continuous Flow
CLW	Costs of Losses and Waste
CPM	Company Productivity Mission
CPS	Company Productivity Strategy
CPU	Capacity Planned Utilization
CPV	Company Productivity Vision
CSS	Cost-Saving Strategy
DCM	Daily Cost Management
DMCIP	Daily Manufacturing Cost Improvement Process
DMIs	Daily Management Indicators
DPPM	Defective Parts Per Million
ECRS	Elimination (E), Combination (C), Rearrangement (R), Simplification (S)
FIFO	First In, First Out
KKIs	*Kaizen* and *Kaikaku* Indicators

KPIs	Key Performance Indicators
LCC	Life Cycle Cost
LWSFA	Losses and Waste Stratification Flow Analysis
MB	Management Branding
MBPS	Multiannual Basic Productivity Strategies
MMCIP	Management by Manufacturing Cost Improvement Policy
MCI	Manufacturing Cost Improvement
MCICP	Manufacturing Cost Improvement Catchball Process
MCPa	Manufacturing Cost Policy analysis
MCPd	Manufacturing Cost Policy development
MCPm	Manufacturing Cost Policy management
MDC	Method Design Concept
MLA	Material Loss Analysis
MMIB	Multiannual Manufacturing Improvement Budgets
MPS	Multiannual Profit Strategies
NCU	Normal Capacity Utilization
NVAA	Non-Value-Added Activities
NVACU	Non-Value-Added Cost per Unit
OEE	Overall Equipment Effectiveness
OLE	Overall Line Effectiveness
OMIs	Overall Management Indicators
OPF	One-Piece-Flow
PBM	Productivity Business Model
PCBG	Productivity Core Business Goals
PDCA cycle	Plan–Do–Check–Act cycle
PFC	Product Family Cost
PL	Physical Loss
PLW	Physical Loss and Waste
PCNG	Productivity Core Business Goals
PMP	Productivity Master Plan
PNIS	Product Number Increase Strategy
PPC	Practical Production Capacity
PST	Problem-Solving Techniques
ROI	Return on Investment
SAS	Speed and Agility Strategy
SDI	Strategic Direction of Improvement
SKPMP	Strategic Key Points on Manufacturing Process
SKU	Stock-Keeping Unit
SOP	Standard Operating Procedure

TBM	Time-Based Maintenance
TOCR	Total Offer for Cost Reduction
TPC	Theoretical Production Capacity
TRL	Time-Related Loss
TtM	Time-to-Market
TtSM	Time-to-Start Manufacturing
VACU	Value-Added Costs per Unit
WIP	Work In Process
WIPS	WIP from Set-up
WIPT	WIP from Transfer
ZCLW	Zero Costs of Losses and Waste

Introduction

After completing my second doctoral thesis in industrial engineering in September 2015, where I developed the concept of the MCPD system using the *action research method*, my first book on the MCPD system entitled *Manufacturing Cost Policy Deployment (MCPD) and Methods Design Concept (MDC): The Path to Competitiveness* was published. This first book on MCPD was written together with Shigeyasu Sakamoto who presented his concept of *Methods Design Concept (MDC)* therein; it had as its main objective the presentation of the new concept of MCPD. My second book on the MCPD system, *Manufacturing Cost Policy Deployment (MCPD) Transformation: Uncovering Hidden Reserves of Profitability* was aimed at presenting how to transform manufacturing companies through MCPD, transformation driven by the need to fulfill external (from sales) and internal (from manufacturing cost improvement or MCI) annual and multiannual manufacturing target profits through uncovering hidden reserves of profitability (CCLW continuous reduction to achieve the annual MCI goal). From the publication of the two books by Productivity Press, the first book in April 2017 and the second book in February 2018, I received many questions from readers, namely from top management, middle management, and professional support staff, about setting out in detail the annual and multiannual MCI targets and means and how to choose and especially implement the most profitable improvement projects to meet the annual MCI goal. So I decided to complete the *MCPD system trilogy* with this new book and to develop how to set MCI targets and means and to choose and implement *kaizen* and *kaikaku* projects for MCI in order to meet the annual MCI goal.

THE BASICS OF THE MCPD SYSTEM

From the perspective of MCPD, the annual and multiannual manufacturing target profits through productivity represents a predetermined stake, a stake of the strategic reduction of CLW and CCLW. Typically, CLW accounts for 30%–40% of the total cost of a manufacturing company and we seek to target productivity improvements as close to the *ideal cost or zero CLW* as possible for each PFC and for the whole company.

The MCPD system approach in this book, like any other approach to the MCPD system, is a structured approach, following the same three phases and seven steps, both at the company level and each PFC level, and at the level of each *kaizen* and *kaikaku* project for MCI.

"The concept of Manufacturing Cost Policy Deployment (MCPD) is defined as the process of translating the strategic objective of reducing manufacturing costs in the long run toward the improvement of annual systematic activities and toward annual systemic improvement actions by setting targets and means to improve process costs of families of products, in order to

- achieve annual MCI targets and means setting for all the main processes of all product family cost (PFC).
- fulfill the annual manufacturing improvement budgets (AMIB; both existing products and new products) by continuous investigation of the relationships between costs, processes, and losses and waste.
- achieve performance of annual manufacturing cash improvement budget (AMCIB).
- direct and plan the systematic and systemic annual manufacturing improvements through continuous reconciliation between the need to reduce costs and process-level opportunities for MCI.
- engage the workforce to meet annual MCI targets and means.
- measure and analyze performance for MCI.
- achieve cost targets at shop floor level. (Posteucă, 2015, p. 65; Posteucă and Sakamoto, 2017, pp. 81–82; Posteucă, 2018, pp. 10–11)

The tangible effects of applying the MCPD system can be found both at the level of each PFC process and at the P&L level due to the development and use of annual manufacturing improvement budgets (AMIB) for existing and future products and at the level of cash flow statement due to the development and use of annual manufacturing cash improvement budget (AMCIB).

WHY MCPD PROFITABILITY SCENARIOS?

The scenarios of annual and multiannual manufacturing target profit fulfillment are alternatives to addressing the annual MCI targets and means fulfillment by continually improving the productivity of each PFC process based on increasing sales trends and the predominant need to maximize outputs ("E") or based on the declining trend of sales and the preponderant need to minimize inputs ("I"). So, the profitability scenarios hypostases (boundaries for the scenario) are predominantly for maximizing outputs ("E") or predominantly for minimizing inputs ("I").

The basic logic of the scenarios for MCI is: product unit cost of manufacturing tends to decrease proportionally with the losses and waste from processes and implicitly CLW. The more and better CLW is identified and addressed along the main processes of each PFC, depending on the profitability scenarios hypostases, the more likely they are to set and meet annual and multiannual MCI targets through MCI means.

From the perspective of the MCPD system, the two profitability scenarios hypostases are linked to the four cases of sales scenarios:

- the first two scenarios of sales (scenario 1: major and flexible sales growth; scenario 2: incremental sales growth) aim at focusing on fulfilling the multiannual and annual manufacturing target profit, especially through external manufacturing profit through maximizing outputs ("E")—the predominant need for productivity growth by improving effectiveness (reducing losses—not effectively used input);
- the following two sales scenarios (scenario 3: incremental sales declines; scenario 4: major unexpected sales declines) aim at focusing on multiannual and annual manufacturing target profit fulfillment, particularly through internal manufacturing profit through minimizing inputs ("I")—the predominant need for productivity increase through improving efficiency (waste reduction–excess amount of input).

For the 2 profitability scenarios hypostases, the maximization of outputs ("E") or the minimization of inputs ("I"), in the book are developed in the eight change drivers (CD) of MCI and system elements for each change drivers of MCI for each PFC as follows:

- for external manufacturing profit through maximizing outputs ("E"):
  - effectiveness of current equipment (CD1);
  - effectiveness of new equipment (CD2);
  - development of new profitable products (CD3);

- for internal manufacturing profit through minimizing inputs ("I"):
  - maximizing variable cost efficiency (CD4);
  - maximizing fixed cost efficiency (CD5);
  - continuously improving manufacturing lead time (CD6);
  - continuously aligning processes to market needs (CD7);
  - continually improving inventory levels (CD8).

In this context, setting both annual MCI targets and annual MCI means (*kaizen* and *kaikaku* for MCI) aim to achieve the annual MCI goal by accentuating one of the two profitability scenarios hypostases.

CHALLENGES OF SYSTEMATIC AND SYSTEMIC IMPROVEMENT OF MANUFACTURING COSTS

The greatest challenge of the MCPD system is the scientific establishment of the annual MCI targets and means (the second step of the MCPD system) to meet the annual MCI goal.

In order to establish annual MCI targets and means for each main process of each PFC and/or for each representative product of each PFC, reconciliation is needed between the need to reduce the product unit costs for the manufacturing stage and the acceptable future status of CLW for the next five years or total offer for cost reduction (TOCR) in order to reach the ideal cost (zero CLW).

Therefore, a continuous and consistent reconciliation between the top-down approach and the bottom-up approach is required to set the annual MCI goal for each PFC by developing seven matrices:

- Matrix 1: continuous measurement of losses and waste
- Matrix 2: analyze and identify sources that generate losses and waste
- Matrix 3: continually convert losses and waste into manufacturing costs
- Matrix 4: analyze and establish critical costs of losses and waste
- Matrix 5: current assumptions for critical costs of losses and waste improvement
- Matrix 6: setting annual MCI targets to achieve annual MCI goal
- Matrix 7: setting annual MCI means to achieve annual MCI targets

With the help of these seven matrices, it is possible to establish profitable annual improvement projects that will contribute to the full accomplishment of the annual MCI goal.

HOW THIS BOOK IS ORGANIZED

To address the two main directions of the MCPD system approach of this book, (1) the scientific establishment of MCI targets and means based on profitability scenarios hypostases and the seven matrices and (2) the successful implementation of profitable productivity improvement projects following the seven steps of the MCPD system continuously, the book is structured in three parts.

This book is divided into three parts and eight chapters, which are described in the following sections.

Part I: Understanding of the MCPD system

The first part of the book presents the role of the MCPD system from the perspective of the directions for increasing the manufacturing profit on the account of productivity in the two possible hypotheses of sales (growth or reduction) through the development of: (1) MCI change drivers; (2) system elements of change drivers of MCI for each PFC; (3) current level of cost of losses and waste (CLW) for each system element of change drivers of MCI; (4) the basic MCI strategies for each current level of CLW; (5) current assumptions for each MCI strategy and (6) MCI targets and means.

At the same time, this first part of the book presents the MCPD system from the perspective of transforming manufacturing companies by establishing and implementing profitable improvement projects. For this purpose, in this first part of the book, the three phases and seven steps of the MCPD system are presented, and emphasis is placed on establishing annual MCI targets and means through the detailed presentation of the seven matrices for setting annual MCI targets and means, and a real example of a manufacturing flow transformation of a PFC through the MCPD system in order to achieve at least 6% annual MCI is presented.

Chapter 1: Let's Start by Really Tackling Improvement of the Current Level of Manufacturing Profit through Productivity Improvement

In this first chapter we present the role of productivity in supporting the multiannual and annual manufacturing target profits and the basics of the MCPD system. Therefore, it presents: (1) the directions for consistent growth of the manufacturing profit and of the role of productivity; (2) the need for productivity-centered thinking to achieve multiannual manufacturing target profits; (3) the productivity support for achieving multiannual manufacturing target profits by increasing the volume of products made (continuous improvement/capacity increase; effectiveness) and sold products (continuous improvement/increase in delivery speed; efficiency) and the reduction of costs associated with goods manufacturing and delivery; (4) the method of sustaining the manufacturing profit through productivity in terms of increasing or decreasing in sales volumes. The MCPD concept is then defined and the basics of the MCPD system are presented to support annual and multiannual manufacturing target profits by meeting MCI targets and means, or through *management by MCI targets and means* or *management by MCI policy* (MMCIP) for a consistent productivity management coordinated by the need for MCI.

Chapter 2: Scenarios and Strategies of the MCPD System

In this second chapter we present the connections between the external manufacturing target profits through maximizing outputs and the internal manufacturing target profits through minimizing inputs with the four potential scenarios of sales volumes evolution to create the prerequisites for how to set business expectations and productivity strategies to meet MCI targets for each PFC, depending on the current and future business context. The MCI change drivers for each PFC and the elements that through successive or systemic upgrades can generate significant beneficial changes to the manufacturing system and the MCI in the four sales scenarios for both external manufacturing profits and internal manufacturing profits are then presented. At the same time, we present the system elements for each MCI change driver for each important process of each PFC to identify the elements and areas where a "magnifying glass" is needed to continuously identify and measure losses and waste and associated CLW in order to establish basic MCI strategies. Further, we present the necessities and directions for continuous scanning of each important process of each PFC to determine the current total level of CLW for each system element of the MCI change drivers, and how to develop the annual and multiannual strategies of MCI by addressing the current level of CLW for each process of each PFC in order to further establish MCI targets and means. Identifying current assumptions for CCLW and defining the most credible assumptions for MCI strategies and presenting how to approach MCI targets and means based on current assumptions for CCLW are the essential ingredients for obtaining the tangible and intangible effects of the MCPD system based on the development of structured scenarios and strategies to meet MCI targets by continuously and harmoniously transforming the manufacturing flow of each PFC.

Chapter 3: The MCPD Transformation: Establishing and Implementing Profitable Improvement Projects

In this chapter, the mechanism for choosing the most profitable productivity improvement projects will be emphasized by steering them in line with the need for a consistent MCI fulfillment. The basis of this mechanism is to establish and meet annual MCI targets, aiming at achieving a continuous and harmonious transformation of the entire manufacturing flow of each PFC and the entire company, depending on the market signals, according to the need for manufacturing profit and depending on the current state of the company's main processes and equipment.

Therefore, this chapter presents the main ingredients for selecting profitable improvement projects. The three phases and seven steps of the MCPD system are presented. The first two steps of the MCPD system will be addressed in more detail to show how MCI targets and means (MCI policy) are established and also how to choose the most profitable improvement projects. At the same time, it will address how to implement profitable productivity improvement projects to meet annual MCI targets.

Chapter 4: Annual Transformation Example through MCPD System: MCI with at least 6% per year for a PFC

In this chapter a real application of the MCPD system in a manufacturing company (electronics company/repeat lot–manufacturing and assembling industry) is presented. In this application, the main actions and activities necessary for a harmonious transformation of the manufacturing flow and the way of addressing the main challenges regarding the consistent implementation of the MCPD system are presented.

It starts by presenting the company background and its current business needs and then shows how to implement the three phases and the seven steps of the MCPD system to successfully achieve MCI of at least 6% per year for the next five years (at the P&L level), amid a manufacturing profit margin of between 30%–40% of total current expenses (CLW current level), a reserve that is exploited annually by MCI means (or *kaizen* and *kaikaku* for MCI), and is based on critical cost of losses and waste (CCLW) and on the development of the critical to annual manufacturing profitability tree (CAMPT). It also presents the tangible and intangible general results of applying the MCPD system to the respective company.

Part II: Profitable Improvement Projects Implementation through MCPD for MCI toward Maximizing Outputs

In the second part of the book, we present real implementations of profitable improvement projects through the MCPD system in different industries (the process industry and manufacturing and assembly), namely the implementation of six *kaizen* projects for MCI and two *kaikaku* projects for MCI, in situations with the need to maximize outputs amid the growth of sales volumes and the need to increase effectiveness and reduce losses (not effectively used input), starting from annual CCLW and from CAMPT.

Therefore, in this second part, real systematic (*kaizen*) or systemic (*kaikaku*) improvement projects will be presented, or other MCI means to meet MCI targets on time, based on credible and capable assumptions for each MCI strategy for change drivers (CD) of MCI to ensure the external manufacturing target profit: effectiveness of current equipment (CD1); effectiveness of new profitable products (CD2); and effectiveness of new equipment (CD3).

Chapter 5: MCI and Increasing Equipment Effectiveness

The continuous improvement of current equipment effectiveness is the core concern of the MCI change driver on the need to increase equipment effectiveness to ensure the external manufacturing target profit. This chapter presents real improvement projects or annual MCI means to meet annual MCI targets converging to annual MCI goal (and CAMPT), by enhancing or eliminating CCLW, in the current order of importance of equipment improvements (quality, performance/speed and availability) to achieve *increasing equipment effectiveness*. First, we present the approach and the results of two real *kaizen* projects to meet MCI targets by improving or eliminating CCLW related to equipment quality; respectively, a *kaizen* project for equipment scrap reduction and one for equipment rework reduction. Then there are two real *kaizen* projects to meet MCI targets by improving or eliminating CCLW related to equipment performance (speed), a *kaizen* project for equipment cycle time reduction and a *kaizen* project for people cycle time reduction (eliminating bottleneck). Finally, there are two real *kaizen* projects to meet MCI targets by improving or eliminating CCLW related to equipment availability, a *kaizen* project to eliminate equipment breakdown and a *kaizen* project for equipment setup, settings and adjustment time reduction.

Chapter 6: MCI and Increasing Manufacturing Capacity through New Products and New Equipment

The continuous improvement of company effectiveness to increase the use of current capabilities through the continuous development of new profitable products and the further increase of equipment capacity through the acquisition of new equipment is a basic concern of MCI change drivers of the need to increase manufacturing capacity (maximizing outputs "E") to ensure external manufacturing target profit.

In this chapter, two systemic (*kaikaku*) improvement projects were presented. There is a project to develop new profitable products to fit the required time-to-market (TtM) and a project to install new equipment to fit the time-to-start manufacturing (TtSM), both projects being converged to the annual MCI goal (and CAMPT).

Part III: Profitable Improvement Projects Implementation through MCPD for MCI toward Minimizing Inputs

In the third part of the book, we continue to present real implementations of profitable improvement projects through the MCPD system in different industries, but this time in a situation involving the necessity of minimizing inputs due to the reduction of sales volumes, and implicitly of the need to increase efficiency growth and reduce waste (excess amount of input).

In this third and last part of the book, there are presented real systematic (*kaizen*) and systemic (*kaikaku*) improvement projects or other MCI means to meet MCI targets on time, based on credible and capable assumptions for each MCI strategy for change drivers (CD) of MCI to ensure internal manufacturing profit (minimizing inputs "I") by: maximizing variable cost efficiency (CD4); maximizing fixed cost efficiency (CD5); continuously improving manufacturing lead time (CD6); continuously aligning processes to market needs (CD7) and continually improving inventory levels (CD8). More specifically, it presents the implementation of three *kaikaku* projects for MCI and four *kaizen* projects for MCI, starting from the annual CCLW and from CAMPT to achieve the annual MCI goal.

Chapter 7: MCI through Improving Variable and Fixed Costs: Material, Environment and Labor Costs

The continuous improvement of company efficiency through continuous improvement of variable and fixed costs is a basic concern of MCI change drivers to achieve *manufacturing control* (minimizing inputs "I"). So, after a manufacturing company has made certain that it has developed the most profitable products (CD2), that it has installed the most suitable equipment (CD3) and that this equipment has an appropriate level of effectiveness at the level of planned production (CD1), it will need to ensure strict control over all the resources consumed throughout the manufacturing flow of each PFC (CD4 and CD5). This chapter presents real systemic improvement projects (*kaikaku*) or annual MCI to achieve *improving variable and fixed costs* (material, environment and labor costs). Therefore, we presented the approach and the results of three *kaikaku* projects for MCI (systemic improvement) by (1) replacing raw material for an existing product to reduce variable costs with materials; (2) reducing environment costs to reduce relative fixed costs with environment; and (3) implementing a simple automation device to reduce fixed costs with direct labor.

Chapter 8: MCI through Aligning the Factory Lead Time and Cycle Time to Takt Time

The continuous improvement of company efficiency through continuous improvement the alignment of manufacturing lead time with takt time is a basic concern of MCI change drivers on the need to increase manufacturing delivery (minimizing inputs "I") to support external manufacturing target profit and especially internal manufacturing target profits. Continuous knowledge of LW and CLW levels related to factory lead time—more specifically, related to supply chain lead time (waste), manufacturing lead time (waste) and delivery lead time (waste)—and continuous reduction of CCLW by planning and running on time the required annual MCI means is required. Thus, in this chapter, four *kaizen* projects for MCI (systematic improvement) were presented by (1) reducing the lead time of an assembly line; (2) increasing the synchronization level of an equipment lead time to an assembly line takt time; (3) reduction of WIP storage space; and (4) reducing unplanned packaging consumption and reduction/ elimination of supply lead time.

The book ends with the presentation of MCPD transformation checklist (Appendix 1) and the MCPD glossary (Appendix 2).

In conclusion, I am confident that the MCPD system will help your company to meet multiannual and annual profit plans by choosing and implementing the most profitable MCI means (systemic and systemic improvement projects for MCI) with the continued and consistent increase in productivity.

Alin Posteucă

Part I

Understanding of the MCPD System

1

Let's Start by Really Tackling Improvement of the Current Level of Manufacturing Profit through Productivity Improvement

Manufacturing companies survive and thrive by manufacturing and selling profitable goods. The manner in which goods are made and making profit differs from one company to another and/or within a company from one period to another. These fluctuations depend on the health of the structure of each manufacturing company for a certain period of time or, more precisely, on their ability to generate a continuous profit, even if there are larger or smaller changes in the internal environment and, especially, the external environment in which they carry out their activity. Often the financial reality is more readily understandable by managers than the reality of manufacturing flow processes. In this respect, it is necessary to disclose directly the health of the manufacturing flow or, more precisely, the current and future level of non-productivity in the company's financial statements (pro-forma and for a completed period).

In this first chapter, we will present the role of productivity in supporting multiannual and annual manufacturing target profit (Sections 1.1, 1.2, 1.3, and 1.4) and the basics of the MCPD system (Section 1.5). In this context, Section 1.1 presents the directions of consistent growth of manufacturing profit and productivity role. Section 1.2 addresses the need for productivity-centered thinking to achieve multiannual manufacturing target profit by manufacturing companies. In Section 1.3, productivity support is shown to achieve multiannual manufacturing target profit by increasing the volume of products manufactured (continuous improvement/capacity increase; effectiveness) and of products sold (continuous improvement/increasing delivery speed; efficiency) and reducing costs associated with the production and delivery of goods. Section 1.4 describes how to support manufacturing profit through productivity while increasing or decreasing sales volumes. Subsequently, Section 1.5 defines the MCPD concept and presents the basics of the MCPD system to support the annual and multiannual manufacturing target profit by achieving manufacturing cost improvement (MCI) targets and means or, in other words, by management of MCI targets and means, or management of MCI policy for consistent productivity management. Section 1.6 sets out the conclusions of this first chapter.

1.1 DIRECTIONS FOR MANUFACTURING PROFIT GROWTH AND PRODUCTIVITY ROLE

Manufacturing companies are under continuous strategic transformation to fulfill their vision and to be able to continuously adapt to the challenges of the internal and especially the external environment. Often the vision of the manufacturing companies is strategic positioning in the market or, more precisely, competitiveness through product volumes destined for specific markets, and viable and consistent profitability in the medium and long term. Therefore, profitability and gaining a certain percentage of the market for the next 5–10 years or more are the core concerns of top managers in manufacturing companies.

In this respect, ensuring the acceptable health of the manufacturing companies structure determines the continuous definition, monitoring, and understanding of the key elements that contribute to the definition and achievement of the multiannual manufacturing profit plan and of the associated target product volumes, respectively:

- To the company's external elements: (1) the price and (2) the number of products actually or potentially requested by customers throughout their life cycle and on each market; and
- To internal elements: (1) acceptable levels of product unit manufacturing cost; (2) acceptable productivity levels of current and future capacity; (3) production regimes (continuous, batch, repeated lot, manual, one-off) and delivery speeds; (4) environmental protection and innovation; (5) increasing quality product ratio and (6) employee motivation–safety and health at work, morale, and lifelong learning.

Starting from the current state of these external and internal multiannual elements, top managers often set strategic goals or managerial expectations of the manufacturing system for these elements (increases or decreases) and future critical directions to be achieved by setting annual targets and means (policy) (Akao, 1991, p. 5) for these detailed elements up to the product family and/or individual product process levels.

The continuous redesign of the vision of manufacturing companies and, furthermore, of the need to strategically transform the flows of the product families and their main processes has as a starting point the definition and realization of the multiannual profit plan and, especially, of the manufacturing profit.

The usual or possible profit growth directions are focused on:

1. Increase of the sales price;
2. Increase in sales volumes; and
3. Unit cost decrease (especially of the product unit manufacturing cost for current and future products).

The simplest way to increase profit by increasing revenue is to increase the sales price. However, for most industries, the sales price is often influenced decisively by the market and, in particular, by the prices of competitors, and it is difficult to obtain profit from successive increases in sales prices without reducing the number of products sold (accepted by customers), at least for certain markets or periods.

Increasing sales volumes could create the potential for greater profit through increased revenue and product unit manufacturing cost reduction by volume, but without a company continually designing and launching profitable new products of acceptable quality, that are unique and possibly revolutionary, it is unlikely that sales volumes would maintain their growth trend in the medium and long term for each target market, as competitors will bring their rival products to the market. Often, before the effective production of new products, new products determine the need for new processes and new technologies that need to be prepared as well as possible, if not perfectly, to limit or eliminate possible future variations in processes, losses variations (not effectively used input) and waste (excess amount of input). Moreover, improving the mix of existing products, even in the context of market globalization, is limited by the number of current and/or potential customers.

Moving forward, ensuring an acceptable and consistently decreasing level of unit cost (especially product unit manufacturing cost) and, implicitly, pricing, is the third way to achieve multiannual profit plan. But the costs and the variation of their targeted reductions depend on how the finished products are made, namely the current methods used in the work processes where the resource inputs are transformed into finished products, at a quality level and delivery time acceptable to customers. Generally, in a "natural" manner, at least in the medium and long term, without focused managerial interventions, most of the costs tend to increase (especially the indirect variables—such as utility costs, costs of raw materials, components and auxiliary materials, some capital costs, costs associated with non-value-added processes, etc.). Some managers may consider these increases as "normal" with many objective justifications. The unit cost reduction targets the reduction of research and development (R&D) costs, marketing costs, general and administrative (G&A) costs, and manufacturing costs. The purpose of this book is to address product unit cost of manufacturing or, more precisely, manufacturing cost improvement by reducing or eliminating unnecessary costs in manufacturing processes.

Therefore, by looking more closely at the current mode of transformation of resource inputs into finished product outputs, it is possible to identify the consumption which is more or less useful, but which can be reduced or eliminated for all categories of inputs: (1) stocks and consumption of raw materials, components, and auxiliary materials; (2) times of work of people and equipment; and (3) utilities that are consumed more or less usefully. Reducing the product unit cost of manufacturing represents the reduction of each cost item in the structure of each product and is achieved by continuously and consistently improving the effectiveness and efficiency of each resource inputted in the company and for each transformation process at the man, machine, method, and material (4 Ms) level, fulfilling exactly the quality level and timely delivery required by customers. By continually and consistently improving effectiveness and efficiency at the level of each manufacturing flow and the main processes associated with them, continuous manufacturing cost improvement (MCI) is sought through both manufacturing costs and initial design costs.

In this respect, in 1963 Peter F. Drucker stated:

> *What is the major problem? It is fundamentally the confusion between effectiveness and efficiency that stands between doing the right things and doing things right. There is surely nothing quite so useless as doing with great efficiency what should not he done at all. Yet our tools—especially our accounting concepts and data—all focus on efficiency. What we need is (1) a way to*

identify the areas of effectiveness (of possible significant results), and (2) a method for concentrating on them.

(Drucker, 1963)

Achieving annual and multiannual MCI targets and means by continually reducing costs of losses and waste (CLW) aims at simultaneously achieving the required levels of both effectiveness and efficiency.

The common denominator of the approach to increasing sales volumes and reducing product unit manufacturing cost is to continuously ensure the required productivity level (capacity and delivery). Increasing sales volume involves increasing customer satisfaction and loyalty and maximizing the use of current and future capabilities or, in other words, maximizing the use of all company assets (maximizing effectiveness, especially equipment effectiveness) and minimizing losses. Unit cost reduction, or, more precisely, reducing product unit manufacturing cost, involves minimizing all resource inputs (minimizing transformation costs, material costs, initial R&D costs and depreciation costs), maximizing efficiency, and minimizing waste).

Quantification of non-productivity in costs (CLW) and the continuous improvement of these costs create the premises for a sustainable assurance of the vision of manufacturing companies—of achieving the sales volumes required and the multiannual and annual manufacturing profit plan. Moreover, the disclosure of these non-productivity costs in the balance sheet and their exact localization at the level of product family processes is what each Board of Directors and each top management team wants in order to have a continuous targeting of all improvements needed to achieve productivity vision.

1.2 PRODUCTIVITY GOAL BASIC THINKING FOR ACHIEVING THE MULTIANNUAL AND ANNUAL MANUFACTURING TARGETS PROFIT

The multiannual and annual manufacturing profit plan requires a continuous synchronization between (1) target product volumes to be achieved and sold (to ensure target market share), and (2) target productivity required. In this context, productivity goal basic thinking aims at conceiving and continuously developing a manufacturing system to ensure:

1. The required speed of manufacturing, delivery and design of new products,
2. An acceptable cost for manufacturing, for the entire supply chain and for designing new products,
3. An acceptable quality, and
4. A creativity and involvement of all people within and outside the company to continuously support culture to improve productivity (including quality).

1.2.1 Implementing the Productivity Business Model with Productivity Policy Deployment and Productivity Master Plan

In order to establish a clear direction for productivity to achieve, in particular, multiannual and annual manufacturing profit plans, top managers teams often develop their own business model aiming to achieve the required productivity, in particular to develop their own productivity business model (PBM) (Posteucă and Sakamoto, 2017, pp. 9–66).

PBM starts from the development of the multiannual productivity vision and mission, and then set productivity core business goals (PCBG) and productivity strategies (PS) associated with PCBG. Based on PS, productivity policy deployment (PPD) is implemented to monitor the implementation of the PS to meet overall management indicators (OMIs) and key performance indicators (KPIs) associated with the OMI with the help of *kaizen* and *kaikaku* indicators (KKIs) and daily management indicators (DMIs). In order to achieve the annual and multiannual KPIs targets, through the support provided by KKIs and DMIs, an annual productivity master plan is developed and updated continuously to ensure the convergence and consistency of all productivity support activities in order to achieve the productivity vision.

In fact, productivity goal basic thinking, the essence of PBM, which often represents company goal-based thinking for manufacturing companies, targets:

- *Meeting company's productivity vision (CPV)* for 3 years, 5–10 years or more. This is the main task of the Board of Directors; it is reviewed at three months and three major concerns are addressed: (1) the main external and internal directions for achieving the multiannual profit; (2) manufacturing target capacity to ensure multiannual target sales volumes for each market/family of products in line with the quantitative competitiveness vision; and (3) multiannual target profit.

- *Definition, awareness and observance of the perennial values of productivity*: (1) respect for the environment (a "green company") and (2) respect for human beings and their happiness and creativity (responsibility for staff morale; full understanding of current tasks; and relieving these tasks through continuous improvements);
- *Meeting company's productivity mission (CPM) for 3–5 years associated with the CPV* is the top management task that is reviewed at 2-month intervals seeking to achieve manufacturing target capacity and annual target profit by continuously developing and monitoring profitability improvement projects (kaizen and kaikaku) through productivity improvement.
- *Defining and detailing core productivity business goals (CPBG)* for the CPM with the help of the *company's productivity strategy (CPS)* is the top management task that is engineered and/or re-engineered on a continuous basis for 1–3 years and involves establishing the methods of:
  a. Manufacturing profit increase;
  b. Increasing the number of products sold (in manufacturing and along the whole supply chain; improving the mix of existing products—developing products perceived as unique by customers; quality increase—increasing customer satisfaction and increasing quality product ratio);
  c. Manufacturing cost decrease (reducing costs behind losses and waste);
  d. Increasing the morale of all employees; and last but not least,
  e. Increasing safety and health.

Manufacturing profit increase involves establishing the multiannual dynamics of annual contributions to external manufacturing profit obtained from goods sales, and internal manufacturing profit from MCI by improving loss-related costs (continuous improvement of current capacities, especially equipment) and waste (continuous improvement of stock levels, especially of work in process (WIP)).

- *Achieving the company's productivity strategy (CPS)* for 3–5 years associated to *CPM* for each family product and furthermore for each department (monthly top management task), especially for:

- *(1) steady increase in sales volumes*: the development of scenarios (descriptions of plausible future alternatives) and strategies (the way of fulfilling a desired future, credible hypothesized scenario, by planning and mobilizing all necessary resources efficiently and effectively) such as by increasing sales in certain markets of different geographical regions or countries; increasing customer satisfaction and loyalty; increasing the value of the company's assets; increasing current and future manufacturing capacities; increased synchronization between manufacturing flow and supply chain; increasing manufacturing flexibility; increasing product quality; reducing the time-to-market of new products and supporting product range increase; increasing people's happiness; increasing people's creativity and innovation; environmental protection/sustainability, social involvement, etc.
- *(2) decrease* of product unit cost of manufacturing; developing scenarios and strategies such as: decreasing material costs; decreasing transformation costs; decreasing design cost; decreasing depreciation costs.

  Therefore, *PCBG* and *CPS* aim to tackle stringent problems/constraints of productivity. For example, for *CPS* to increase synchronization between manufacturing flow and supply chain for a particular product family, *PCBGs* can be set up such as: (1) reducing manufacturing lead time for each main process; (2) reducing the number of workstations; (3) reducing the cycle time; (4) reducing the setup and transfer time; (5) reducing the factory lead time; (6) reducing the supply lead time and (7) reducing the delivery lead time.
- Defining, implementing and achieving productivity policy deployment (PPD), or productivity targets and means deployment (Akao, 1991, p. 5), to continuously support *PCBG* and *CPS* for the next 1–3 years. It starts from the definition of targets and means for the Overall Management Indicators (OMIs), and further on, targets and means for each of the OMIs-related Key Performance Indicators (KPIs) by developing and implementing targets and means for KKIs and DMIs based on the annual and multiannual productivity master plan (PMP).

PMP performance is the ongoing task of all people within and outside the company, for all products and for all main processes of each PFC.

1.2.2 The Two Directions of Achieving Manufacturing Target Profit through Productivity

Getting back to the prime rationale of existence of any manufacturing company—i.e. to achieve a reasonable level of multiannual profits—it is attained by establishing and achieving annual target profit at the level of the entire manufacturing company, at the level of each product family, and/or the level of product summed for the total life cycle. By allocating the multiannual target profit on the necessary contribution of the annual target profit for each year, the method of achieving the vision on an annual basis is determined. The continued development of a solid PBM and the proactive pursuit of means of meeting targets for IMOs and, implicitly, for KPIs at process level constitute the big challenges for manufacturing companies. Any variation in meeting targets at predefined deadlines is the subject of rapid problem-solving interventions.

In this context, the strategic reasoning at the product level, at least in the conception and development phase, is often that the target profit is extracted from the target sales price and thus the acceptable level of costs or target costs is obtained. Extending this reasoning of target setting from the operational perspective of all product families and all products of a manufacturing company, starting from the current and multiannual target state converging with CPV, annual target profit can be set as follows:

$$atP = (atSP - atMC)^* atNUS \qquad (1.1)$$

where:

    atP   = Annual Target Profit
    atSP   = Annual Target Selling Price
    atMC = Annual Target Manufacturing Cost
    atNUS = Annual Target Number of Units to Be Sold

While the *atSP* level (as an external entity of manufacturing companies; imposed by competitors and customers, and determined by marketing department analysis as an annual average) cannot be influenced too much to contribute to achieving *atP* (as an external entity of manufacturing companies; imposed by shareholders and competitors), then the *atMC* level (or acceptable target level of efficiency) and the *atNUS* level (or acceptable target level of effectiveness) are the two areas in which to intervene by developing profitability strategies and scenarios in the short, medium and especially the long term, and implicitly through the development of scenarios and productivity strategies associated with these profitability scenarios.

The goal of developing the associated profitability and productivity scenarios is to defuse relationships in establishing policy deployment or between targets and means for OMIs and KPIs to meet CPV and CPM. In this way, the *atP* from the entire manufacturing company translates into the main processes of the product families and the main processes of each product, taking into account the need to achieve the multiannual profit or profit goals, and the life cycle of products and technology.

Even though this rationale for *atP* scenarios and strategies extends to the entire supply chain, we will focus on the manufacturing area of the manufacturing companies.

In this context, the *atP* level for a manufacturing company is determined taking into account:

- *External factors* such as: the level of profits unrealized in previous years according to the multiannual profits plan, the volumes accepted by the market for each type of existing and future product of the company; the adherence to an acceptable level of the current and future price; the validity over time of the planned life cycle of the existing and future products, etc.;
- *Internal factors* such as: ensuring the actual manufacturing capacities needed to meet customer demand for current and future products, or synchronizing manufacturing to market needs to ensure timely deliveries; ensuring an acceptable level of product unit cost of manufacturing; ensuring acceptable manufacturing flexibility; ensuring an acceptable level of product quality; ensuring an in-work environment without risk of accidents and/or occupational illnesses and with a high level of employee morale and creativity; future product placement in the time-to-market (TtM) predefined; future equipment placement in time-to-start manufacturing (TtSM) predefined, etc.

All these internal and external factors are directions for the development of profitability strategies and scenarios by ensuring an acceptable level of productivity at the execution level of strategies and, implicitly, ways to reach a predetermined *Return On Investment (ROI)* in dynamics.

In order to ensure an *atMC* it is necessary to focus on *transformation costs* (direct labor costs, indirect labor costs, manufacturing overheads, depreciation costs), *material costs* (direct material costs and indirect material costs), and, in particular, on *costs of losses and waste (CLW)*. The strategic objective of the CLW is to be zero or as close to zero as possible. Zero costs of losses and waste (ZCLW) is a goal achieved by continuous productivity improvement in order to reach in time a structure of a product unit cost of manufacturing as healthy as possible from the perspective of price acceptability by customers. Therefore, the *atMC* level is normally influenced by the current and target level of productivity, including the target product quality level.

1.2.3 External and Internal Manufacturing Profits and Basic Productivity Strategies

To achieve multiannual manufacturing profits or manufacturing profit goals, *multiannual profit strategies (MPS)* are required to be developed through the creation of *multiannual basic productivity strategies (MBPS)* in order to:

- Ensure an acceptable level of effectiveness to achieve the annual and multiannual target number of units to be manufactured and sold (by maximizing outputs and reducing inputs not effectively used–loss improvement);
- Ensure an acceptable level of efficiency to achieve the annual and multiannual manufacturing target cost (by minimizing inputs and reducing excessive amount of input–waste improvement).

Note 1: Losses (especially transformation times): seven equipment losses (failure losses, set-up/adjustment losses, cutting-blade losses, start-up losses, minor stoppage/idling losses, speed loss and defect and rework losses), five human work losses (management losses, motion losses, line organization losses, losses resulting from failure to automate, measuring and adjustment losses), and three manufacturing resource losses (yield losses; energy losses; and die, jig, and tool losses) (Nakajima, 1988; Shirose, 1999, pp. 40–61). Variation and variability in work methods or the output of equipment capacity (*mura*) are found especially at the level of losses.

Note 2: Waste: this means inventory, which refers to material elements (Ohno, 1988, pp. 19–20; Posteucă and Sakamoto, 2017, pp. 28–30; Posteucă, 2018, p. 16). Other activities, apart from those related to the transfer of WIP between the main processes of manufacturing flow which consume resources without creating value for customers (*muda*), are included in losses. Activity-planning (charging) over equipment and human capacity (*muri*) is found especially at inventory level and particularly in WIP associated with process transfer. In fact, from the manufacturing perspective, where the manufacturing flow speed is significantly influenced by equipment effectiveness, the level of inventory (waste) above the acceptable standard is largely an effect of losses or, more precisely, of the uncontrolled variation of outputs of equipment capacity.

Thus, the acceptable level of effectiveness and efficiency depends in particular on ensuring synchronization of current and future manufacturing capacities with current and future market needs, and the continuous identification and reduction of losses and waste and their associated costs from the level of all processes of each product family, and eventually from each product process level through to the implementation of improvement projects to support annual and multiannual manufacturing target profit.

Therefore, from the perspective of MPS and MBPS, it is necessary:

- To develop scenarios and strategies for achieving external sales profit, especially in the context of a growing sales trend (external manufacturing profit through maximizing outputs or effectiveness, especially by reducing or eliminating losses and their associated costs to ensure a healthier structure of transformation costs, including direct and indirect labor costs); and
- To develop scenarios and strategies for achieving internal profit, especially in the context of a downward trend in sales (internal manufacturing profit through minimizing inputs or efficiency, especially by reducing or eliminating waste and associated costs, in order to ensure a healthier structure of material costs).

As can be seen in Figure 1.1, multiannual manufacturing profit is obtained both from external manufacturing profit and from internal manufacturing profit. Manufacturing cost improvement (MCI) plays a decisive role in both situations: supporting external manufacturing profit through cost of loss improvement and, implicitly, by unlocking the potential capabilities of current and/or future equipment behind equipment losses to support the necessary sales volumes; supporting internal manufacturing profit by continuously improving all CLWs to achieve manufacturing unit cost improvement and, implicitly, supporting medium-term to long-term sales).

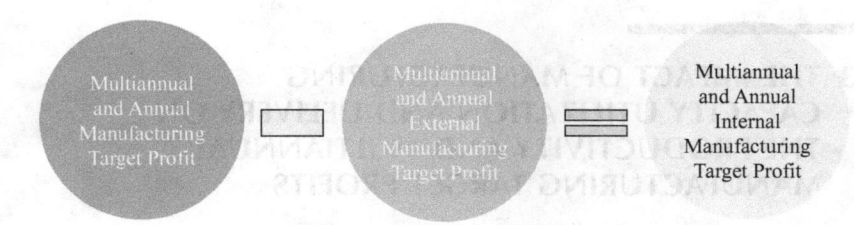

FIGURE 1.1
Multiannual manufacturing target profit equation.

If the multiannual manufacturing target profit is frequently established and communicated, just like multiannual external manufacturing target profit (or profit obtained from the sales of goods and services), then the achievement of the multiannual internal manufacturing target profit is still the basic challenge of many manufacturing companies, or the main unknown element. Mastering the multiannual internal manufacturing target profits ensures increased comfort of mastering ROI by predefined performances of profitable systematic (*kaizen*) and systemic (*kaikaku*) improvement projects to achieve the annual and multiannual CLW targets.

In this context, some questions arise: are there ways to identify and increase the multiannual internal manufacturing profits, and will there be positive effects on achieving the multiannual external manufacturing target profit? Through a scientific approach, it will be demonstrated below that there are significant positive effects of multiannual internal manufacturing target profits on multiannual external manufacturing profit increases by increasing the productivity of the core processes of each product family, by continuously involving all people within and outside the company, based on rigorous strategic planning of systematic and systemic improvements, and by continuous monitoring of the outputs achieved at both managerial and shop floor (operator) levels.

In conclusion, through MPS and MBPS, achieving the multiannual target profits and manufacturing target profit goals should not change regardless of sales volume progress. Specifically, long-term ROI fulfillment in absolute and/or percentage is required to ensure consistent health of manufacturing companies regardless of the trend in sales through consistent productivity. As will be now be shown, this is achieved by setting and fulfilling the annual MCI goals of all CLW for: the annual MCI goal group-wide, the annual MCI goal for each plant, the annual MCI goal for each product family cost, and the annual MCI targets and means for each main process.

1.3 THE IMPACT OF MANUFACTURING CAPACITY UTILIZATION AND DELIVERY ON THE PRODUCTIVITY AND MULTIANNUAL MANUFACTURING TARGET PROFITS

Productivity support is critical in achieving the multiannual manufacturing target profits by increasing the volume of products created (continuous improvement or increase in manufacturing capacity; effectiveness) and by increasing the volume of products delivered and sold (continuous improvement or increase in the speed of delivery; efficiency) combined with the need to continuously reduce the costs associated with the realization and delivery of goods.

Manufacturing capacity utilization or the current capacity to generate outputs by the current manufacturing method depends especially on:

1. The capacity level originally projected;
2. Maintaining the initial capacity level through systematic improvements in order not to result in forced capacity degradation;
3. The performance of manufacturing and supply planning of raw materials, materials and spare parts; and last but not least,
4. The level of market demand for current and future products— implicitly through the inclusion of TtM in future products.

The achievement of atNUS (or the acceptable target level of effectiveness) is based on the design of the synchronization between the required capacity utilization (capacity planned utilization or CPU) and the target manufacturing level imposed by the target volume level of sales.

From the manufacturing perspective, the CPU mainly targets equipment, operators, manufacturing, and storage facilities in manufacturing areas and in warehouses and utilities.

CPU determination takes into account:

- *Theoretical production capacity (TPC)*: this is the maximum capacity that the factory was designed for. The TPC is sometimes called ideal capacity and is often impossible to meet in the actual time available in reality to achieve the production;
- *Practical production capacity (PPC)*: starting from the TPC, this is determined based on historic measurements of downtime for

equipment, operators, and utilities. More precisely, the downtime is subtracted from the TPC. Downtime often includes: average days of delays for receipt of raw materials from suppliers, planned maintenance, average days of absenteeism, average days of medical leave, time for short breaks, time for leave, average days of participation of operators at training, and average days of interruption in electricity supply and so forth. All these periods of downtime, which are subject to systematic (*kaizen*) and systemic (*kaikaku*) improvements, are considered periods that have objective justifications for a given time, but the downtime must decrease over time;

- *Normal capacity utilization (NCU)*: starting from the PPC, the determination of the capacity required to meet the volume of manufacture imposed by customers i.e. the time allocated to the manufacture. The NCU is determined at both multiannual and annual levels. At a multiannual level, it is based on the synchronization between the product life cycle volumes and the life cycle volumes of the equipment in particular). At the annual level, it is based on seasonal periods, when it is necessary to ensure maximum volumes for relatively short periods of time, both for manufacturing and storage capacities. NCU takes into account the planned level of OEE (Overall Equipment Effectiveness) and eventual level of OLE (Overall Line Effectiveness) based on their historical level and on the basis of planned improvements for the following periods (through *kaizen* and *kaikaku*).
- *Capacity planned utilization (CPU)*: this is based on the NCU. The CPU represents the expected use of manufacturing capacities to achieve the planned manufacturing quantity for the target period i.e. the time required to achieve production.

In this context, the CPU can be in one of the following three states:

1. The overcapacity state: *capacity planned utilization (CPU) < normal capacity utilization (NCU)*. A determinant role is the correct setting of *practical production capacity (PPC)*.
2. The undercapacity state: *capacity planned utilization (CPU) > normal capacity utilization (NCU)*. Although the CPU should normally be larger than planned manufacturing volumes (the previous overcapacity) by 5%–20% to cope with unforeseen situations, both external (additional customer requests) and internal (overtaking of

standard downtime and of historic OEE/OLE), it is sometimes lower than the NCU. This is a state of temporary overloading of capacities that, in the short term and especially the medium and long term, leads to the occurrence of losses and especially of waste (inventory) in excess and, implicitly, potential multiannual manufacturing profit erosion. Once again, a decisive role in preventing this is played by the establishment of a correct *practical production capacity (PPC)*.

3. The undercapacity with a hidden overcapacity state. This is the state where the standard downtime level and the historical OEE/OLE are significantly outweighed, or there are no consistent measurements and concerns for systematic and systemic improvement in this respect. In this context, inconsistent workloads are often planned.

Therefore, starting from Ohno's view on capacity ("present capacity = work + waste") (Ohno, 1988, pp. 19–20), ensuring an acceptable level of productivity and profitability requires an approach involving continuous improvement of "work" (man-hours by product and set-up time for main equipment), of losses (Overall Equipment Effectiveness or OEE and Overall Line Effectiveness or OLE), and of waste (total lead time, material inventory days, product stock days, WIP inventory days).

In conclusion, a question arises: How can the systematic (*kaizen*) and systemic (*kaikaku*) improvement projects be scientifically chosen and how are they successfully implemented, taking into account the types of losses and waste in the main processes of the product families and their associated costs, to ensure annual and multiannual manufacturing target profit under overcapacity or undercapacity states? This is the basic question that will be answered in this book.

1.4 IMPROVEMENT OF THE CURRENT LEVEL OF MANUFACTURING PROFIT THROUGH PRODUCTIVITY

Often large manufacturing companies have the majority of methods, techniques, and tools to improve productivity, but their use is not always directed to reaching the manufacturing target profits level. If there were many tools in a sculptor's workshop, one might think it meant that this is a sculptor who manages to delight viewers with his works. It is often not so. A successful sculptor first has in mind exactly what he wants to do, and then he buys and uses all the tools necessary for each phase of his future work. A valuable sculpture work is never a random occurrence. The same is true for manufacturing companies. An important part of operating profit is wasted by the lack of proper and complete use of productivity methods, techniques, and tools, and by the failure to aim continuously at productivity improvements to manufacturing target profits.

Therefore, given the relatively low possibility of intervention on the rise in prices, and with a pressure on their continuous decrease, in order to ensure the fulfillment of annual and especially multiannual manufacturing target profits, manufacturing companies need a consistent approach of the productivity of manufacturing capacities, and delivery of current and future manufacturing products, manufacturing cost improvement (MCI), and cost improvement across the supply chain, with the help of various systematic (*kaizen*) and systemic (*kaikaku*) improvement projects of productivity, irrespective of the fact that:

- Sales volume is increasing for the following periods (ensuring *atP*, especially by *maximizing the effectiveness* of the main product family processes; by accentuating the achievement of *external profit* mainly by manufacturing outputs, especially equipment and maintaining process inputs at least at a constant level; and by not effectively using input or *losses*) or
- Sales volume is decreasing for the following periods (ensuring *atP* especially by maximizing the *efficiency* of the main processes of the product families, especially for raw materials, auxiliary materials and direct labor; by accentuating the achievement of the *internal profit* mainly by minimizing inputs and maintaining the output of processes at least at a constant level; and by limiting or reducing excess amounts of input or *waste*).

We will detail these aspects in the following section.

1.4.1 Nearly Total Dependency of Current Level of Manufacturing Profit on Sales Dynamics

Currently, for many manufacturing companies, the performance level of the multiannual manufacturing profit is overwhelmingly affected by the number of products made and sold (external profit). Figure 1.2 shows an example of the dependence between the manufacturing profit level and

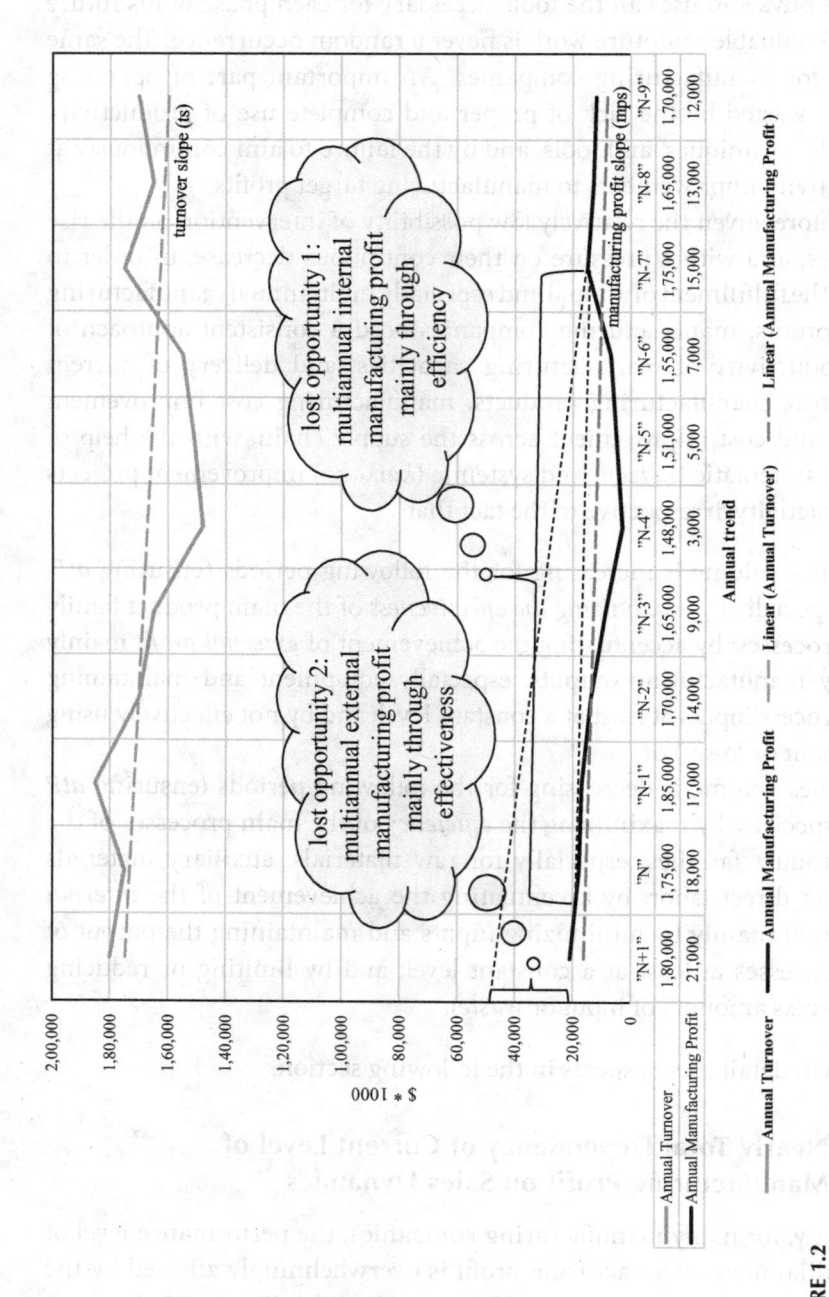

FIGURE 1.2

Annual relationships in dynamics between turnover and manufacturing profit without substantial productivity concerns.

the number of products achieved and sold in a company that has no real and/or consistent concerns to sustain manufacturing profit through productivity ("AA-Plant"–electronics company/repeated lot–manufacturing and assembling industry) and the judicious use of available capacities and deliveries on time.

As can be seen in our example in Figure 1.2, the growth rate of turnover (ts) and manufacturing profit (mps) are relatively small and almost parallel, and the risks of irreparable loss of competitiveness are high. After all, over time, "AA-Plant" missed two types of opportunities:

- Lost opportunity 1: multiannual internal manufacturing profit mainly through efficiency. This missed opportunity mainly concerns the consistent and continuous failure to address resource input minimization (especially through *kaizen* to reduce the consumption of raw materials and labor) amid a constant or declining demand for existing and even future products, the main failure being not maintaining the slightly increasing trend of multiannual manufacturing profit.
- Lost opportunity 2: multiannual external manufacturing profit mainly through effectiveness. This lost opportunity mainly concerns the failure to use at their best the current manufacturing capacities (especially by *kaizen* for the continuous improvement of equipment effectiveness) and the failure to increase new manufacturing capacities (by *kaikaku*: new equipment, new technologies, etc.) amidst an increasing demand for existing products at a given time or new products with a potentially acceptable profit, the main non-achievement being the failure to maintain the increasing trend of multiannual manufacturing profit.

Often, the two opportunities lost are self-determining one another. One of the basic causes of this self-determination is the lack of a managerial approach to profit and productivity in the medium-term and the long-term and, implicitly, overloading current capacities in certain periods, the lack of standardization, and overloading of equipment and people at certain times. Obtaining manufacturing profit through high productivity increases and overloading current short-term capabilities (for one month, several months or even one year) is relatively easy to achieve. The challenge is to achieve medium-term to long-term manufacturing profits on the backdrop of a consistent health of the company's processes and a normal load of capabilities. Therefore, another question arises: how do we set and meet the targets to obtain multiannual manufacturing profits for the core processes of each product family without irreversibly affecting the medium- and long-term capabilities? Since this question is one of the basic questions of this book, it will be answered in the later paragraphs and sections. The main causes of such a situation as shown in the "AA-Plant" are multiple. The following can be explained:

- Target and means for OMIs and furthermore targets and means for each set of OMIs-related KPIs are poorly established. The following situations are often encountered: (1) poor definition at the level of each hierarchical level and at the level of the main processes; (2) definition of KPIs irrelevant for supporting strategies and objectives; (3) KPIs are difficult for managers and the rest of the employees to understand; (4) information is gathered and reported at incorrect time intervals on KPIs (including collection and reporting delays); (5) incompletely gathered information on KPIs; (6) information behind KPIs is easily criticized by managers who need to provide answers for situations of non-performance; (7) unrealistic KPIs targets in relation to the means needed to be implemented through improvements, etc.
- Review and poor analysis of OMIs and KPIs targets and means. The following situations are often encountered: (1) lack of analysis of deviations from targets at regular intervals; (2) lack of a standard of daily and periodic meetings, with clear roles for each participant; (3) lack of a reporting standard for the main 3–5 problems at daily meetings; (4) lack of an interdepartmental approach to deviations; (5) lack of interdepartmental responsibility for solving deviations from established targets; (6) lack of monitoring of a continuously revised action plan for each area of the company and overall company for all relevant deviations; (7) establishing insufficient means to achieve targets and completely, etc.;
- The lack of a consistent methodology of the correct (effectiveness) and complete (efficiency) implementations of the process improvement projects, a continuous methodology connected to the changes originating from outside the company. The following situations may arise: (1) the almost total allocation of all resources to achieve manufacturing and sales, which results in a visible circle of continuous deterioration in the process performance level due to the lack of standardization and

continuous improvement; (2) lack of objective diagnosis of problems and prioritization in addressing problems; (3) lack of support for strategies with actual process data and facts and SWOT analysis (Strengths, Weaknesses, Opportunities, and Threats); (4) postponing the addressing of problems; (5) lack of an interdepartmental approach to problems; (6) lack of clear interconnections of internal metrics (KPIs) with external performance drivers; (7) lack of profound understanding of, or misapplication of, improvement methodology, tools, and techniques due to a lack of awareness of the impact of improvements on operating profit in the medium to long term.

The healthy objective of a manufacturing company is to have a steady growth in manufacturing profit regardless of turnover fluctuations (acceptable turnover fluctuations, rather than switching to completely new product technologies) and limiting "objective" managerial excuses of turnover decrease (economic crises, end-of-life products—those that are obsolete or outdated, rising suppliers' prices, or changing vendor technologies, poorly prepared or fluctuating workforce, obsolete equipment, etc.) Without denying prior negative influences, it is often about the use of an inappropriate labor method and a lack of continuous harmonious transformation of the company's processes according to the market signals.

1.4.2 Improvement of the Current Level of Manufacturing Profit through Productivity

Going forward with an example contrasting with that of "AA-Plant," Figure 1.3 shows an example of the dependence between the manufacturing profit level and the number of products manufactured and sold in a company that has consistent focus on to supporting manufacturing profit through productivity ("BB-Plant"–electronics company/repeat lot–manufacturing and assembling industry). As can be seen in Figure 1.3, "BB-Plant" is starting at the same level of annual turnover as "N-9," but has upward variations in the "AA-Plant" for turnover and manufacturing profit.

The two types of lost manufacturing profit opportunities encountered in the "AA-Plant" example are eliminated in "BB-Plant" by the continuous development of a consistent master plan for long-term productivity regardless of the expected sales volume. As can be seen in Figure 1.3, the multiannual manufacturing profit slope (mps) is increasing irrespective of sales volume (ts). This situation is the desired state of continuous growth

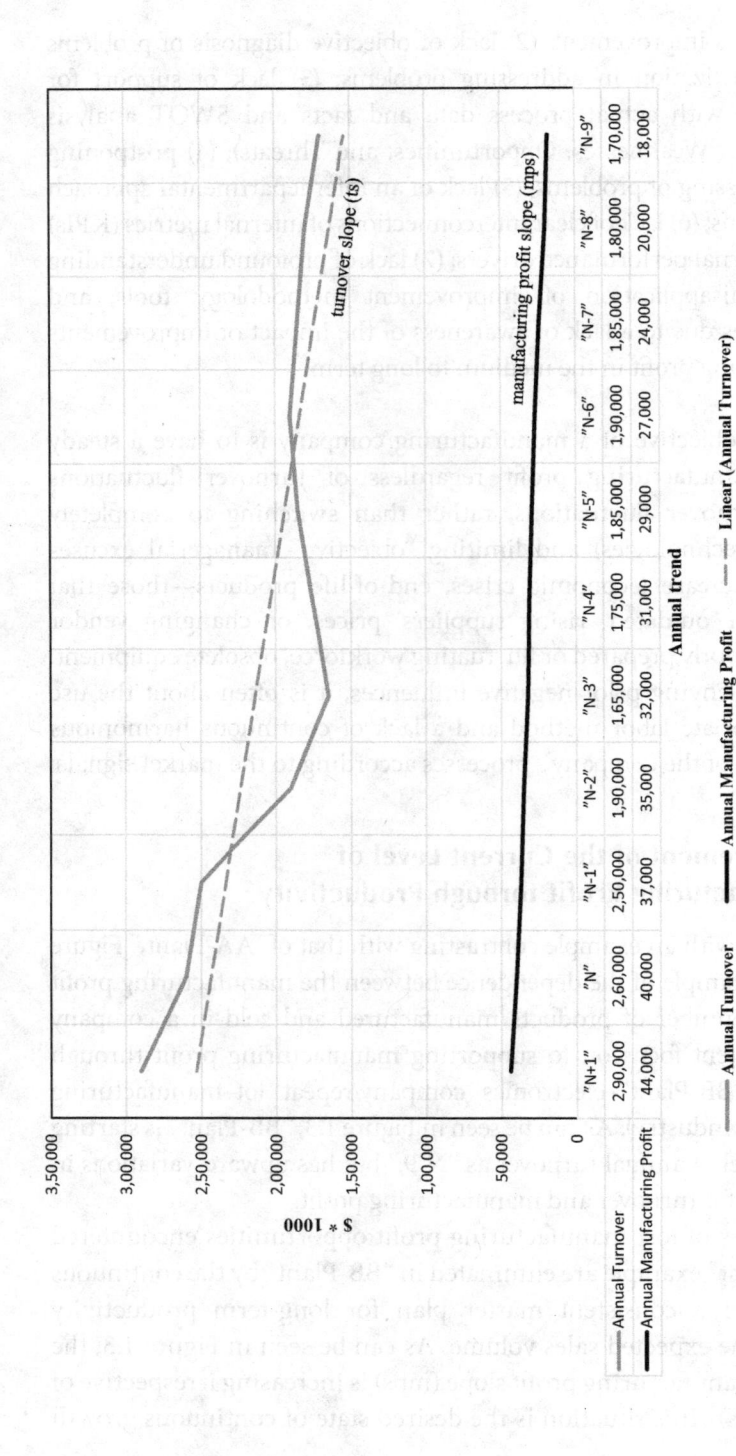

FIGURE 1.3

Annual relationships in dynamics between turnover and manufacturing profit with consistent productivity concerns.

of manufacturing profit with the help of productivity and the judicious use of current and/or future manufacturing capacities.

In the case of "BB-Plant," in order to be able to sustain continuously the volume of planned turnover according to the productivity vision, the necessary manufacturing capacity is planned by the productivity improvement projects, both systematically (*kaizen*) and systemically (*kaikaku*). As can be seen in the example in Figure 1.4, two major types of scenarios are distinguished in "BB-Plant," scenarios intuited and used over time:

- Scenario 1: Maximizing outputs (for the years "N-9"; "N-8"; "N-7"; "N-6"; "N-2"; "N-1"; "N"; "N+1"), especially by reducing or eliminating ineffectively used input (losses), and implicitly by increasing effectiveness, especially for equipment:
  - In the "N-9" years ("I-8"; "N-7"; "N-6"; "N-2"), *kaizen* projects were implemented in particular to support the PN through the CPU. The CPU and PN relations were considered within normal limits; since the CPU was higher than the PN, it was able to cope with unforeseen situations of additional demand from market or temporary in-work incapacity of the company—for example, equipment breakdown or lack of raw materials.
  - In the years "N-1" and "N" it was necessary to implement *kaizen* projects particularly to ensure an acceptable level of the CPU in support of the PN. Relations between the CPU and the PN are considered at the limit. Sometimes, in the short term, the CPU may be less than the PN. In that case, it may sometimes lead to the temporary overloading of the NCU. It is a situation that justifies the need to increase capacity by investing in new capacities to ensure the PN required by customer requests: *kaikaku* project planning and implementation or systemic improvement.
  - In the N year, implementation of *kaikaku* projects in particular as well as *kaizen* projects is required in order to ensure the CPU to support the NP for the year "N + 1" and for the next years according to the productivity vision: quantitative; for competitiveness).
- Scenario 2: Minimizing inputs (for years "N-5"; "N-4"; and "N-3"). In all these years it is necessary to implement *kaizen* projects in particular to limit the excess amount of input (waste) by increasing efficiency, especially for human work, materials and utilities.

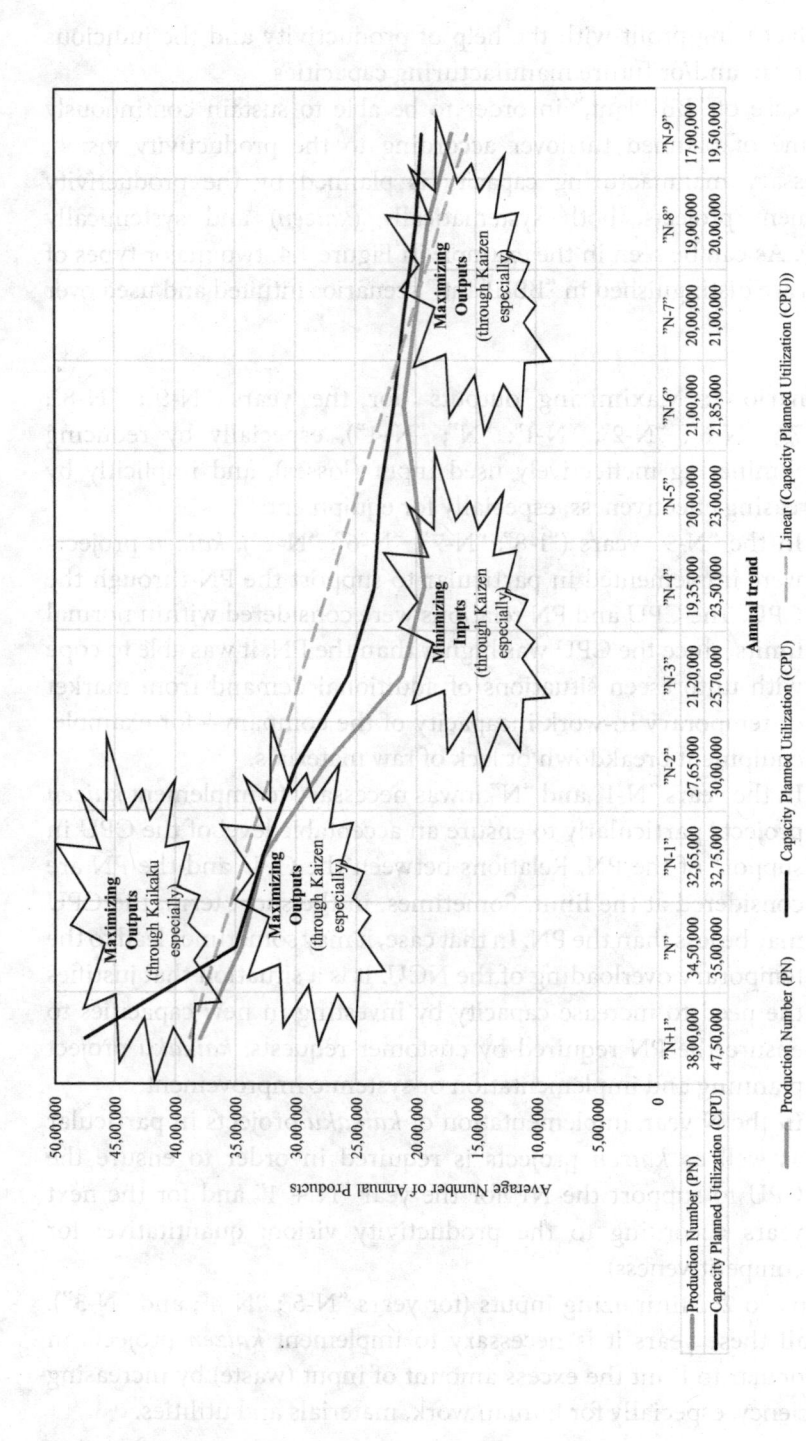

FIGURE 1.4

Dynamics of annual and multiannual scenarios of maximizing outputs or minimizing inputs.

Even though the "BB-Plant" situation is the situation for all product families (the one from Figure 1.4), for certain product families and for some main and/or secondary processes of certain product families, mixed Scenario 1 and Scenario 2 situations can often be encountered, at least temporarily. This is why every *kaizen* project and *kaikaku* project is planned and implemented in every manufacturing company for both types of scenarios.

1.4.3 Establishing Profitability and Productivity Improvement Targets Based on a Dynamic Relationship Between Price, Profit and Cost

Continuing with the "BB-Plant" example, in the context of the need for capacity and timely deliveries, one of the core concerns of the management teams is to ensure an acceptable synchronization between the CPU and the PN and, implicitly, a strategic planning of acceptable extra temporary capacity (the CPU being the reference). Figure 1.5 shows the situation of this acceptable extra temporary capacity over the years for "BB-Plant." As an extension of the vision of the manufacturing company, ensuring the acceptable extra temporary capacity level is necessary in order not to create conditions for subsequent occurrence of waste and losses due to a deficient design of the capacities.

To ensure this acceptable level of extra temporary capacity, often between 10%–20% (these percentages may vary depending on the company's

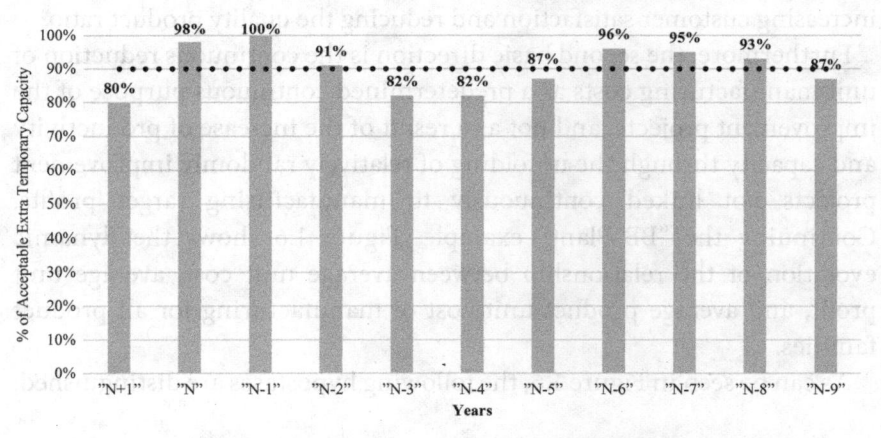

FIGURE 1.5
Strategic evolution of extra temporary capacity.

specific business and business needs for a certain period of time), it takes a synchronization of life cycles of:

1. Current and future products;
2. Equipment;
3. Skill and knowledge levels of operators and especially managers;
4. Capabilities of suppliers of raw materials, materials and logistics; and
5. Warehouses, domestic logistics and external logistics in line with manufacturing volumes.

For example, one of the problems often encountered is the lack of synchronization between the manufacturing capacities of a manufacturing company and manufacturing takt time, with the available manufacturing capacities of raw materials and materials suppliers and supply lead time. In particular, the demand for an apparently lucrative product and a well-planned manufacturing capacity, internal logistics, and storage associated with this demand are all useless if the manufacturing capacities of the suppliers are not available to the company or significant timely supply irregularities occur. Such situations cause systemic losses and waste.

Therefore, the first of the main directions of multiannual manufacturing target profits are to increase the number of products manufactured and sold with the help of productivity increases and to ensure manufacturing capacities for existing and new products through product range increases. Increasing productivity and capabilities are also implicitly based on increasing customer satisfaction and reducing the quality product ratio.

Furthermore, the second basic direction is the continuous reduction of unit manufacturing costs as a predetermined continuous purpose of the improvement projects, and not as a result of the increase of productivity and capacity through the unfolding of relatively randomly improvement projects not linked continuously to manufacturing target profits. Continuing the "BB-Plant" example, Figure 1.6 shows the dynamic evolution of the relationship between average unit cost, average unit profit, and average product unit cost of manufacturing for all product families.

As can be seen in Figure 1.6, the following hypostases are distinguished:

- For Scenario 1 (predominant by maximizing outputs, for "N-9"; "N-8"; "N-7"; "N-6"; "N-5" and "N-4" years) average product unit cost of manufacturing had a steady annual reduction of around 5%

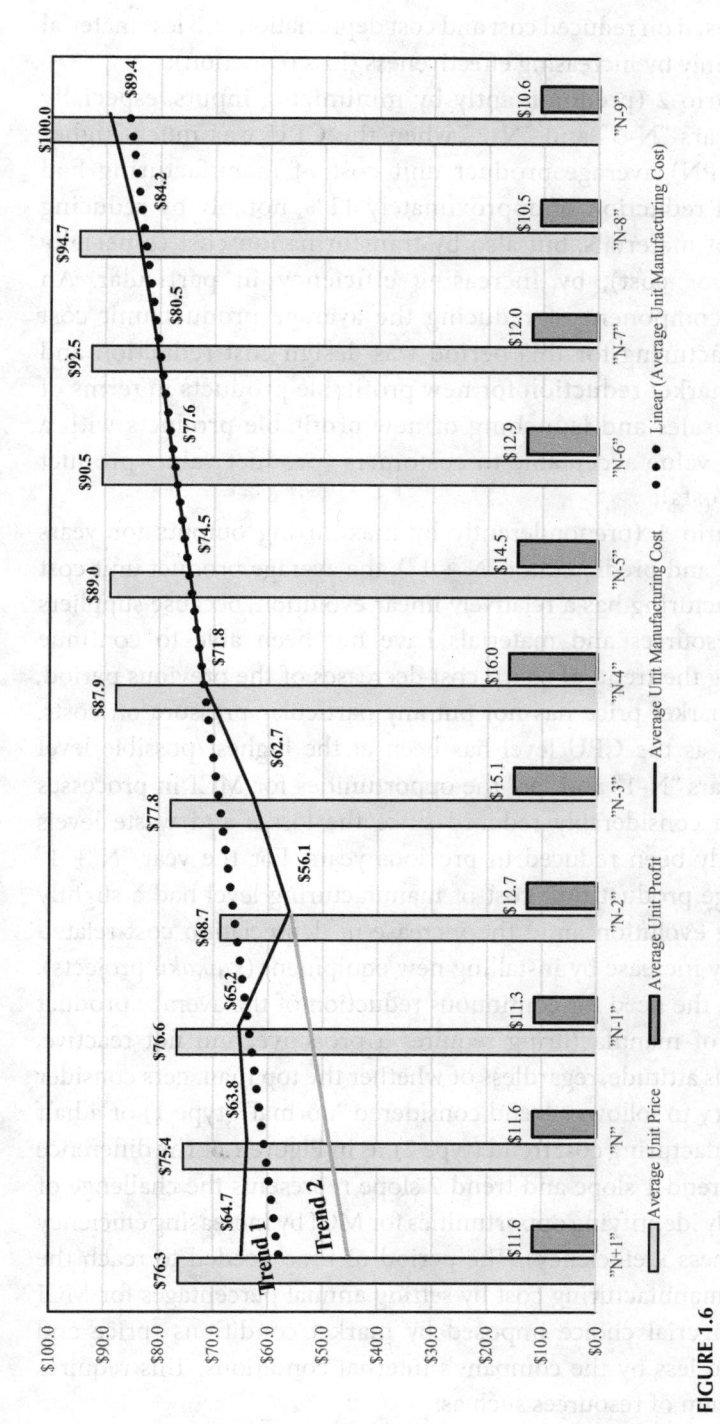

FIGURE 1.6

The dynamics of the relationship between average unit price, average unit profit and average unit manufacturing cost.

(mainly based on reduced cost and cost depreciation and less material costs), mainly by increasing effectiveness (loss reduction);

- For Scenario 2 (predominantly by minimizing inputs, especially for the years "N-3" and "N-2" when the CPU was much higher than the PN) average product unit cost of manufacturing had an annual reduction of approximately 11%, notably by reducing the cost of materials, but also by transformation cost (especially direct labor cost), by increasing efficiency in particular. An essential component of reducing the average product unit cost of manufacturing for this period was design cost reduction and time-to-market reduction for new profitable products in terms of declining sales and launching of new profitable products with a perceived value acceptable to customers (product value=product features/costs);

- For Scenario 1 (preponderantly by maximizing outputs for years "N-1," "N" and predicted for "N + 1"), the average product unit cost of manufacturing has a relatively linear evolution, because suppliers of raw resources and materials have not been able to continue supporting the trend of sharp cost decreases of the previous period, and the market price has not put any particular pressure on costs. Moreover, as the CPU level has been at the highest possible level for the years "N-1" and "N," the opportunities for MCI in processes have been considerably reduced since the losses and waste levels had already been reduced in previous years. For the year "N + 1" the average product unit cost of manufacturing level had a slightly increasing evolution amid the increase in depreciation cost related to capacity increase by installing new equipment (*kaikaku* projects). Therefore, the need for continuous reduction of the average product unit cost of manufacturing requires a proactive, and not reactive, continuous attitude, regardless of whether the top managers consider it necessary to follow: a trend considered "normal" (type 1) or a half unit manufacturing cost trend (type 2), as in Figure 1.6. The difference between trend 1 slope and trend 2 slope represents the challenge of continually identifying opportunities for MCI by increasing efficiency (effectiveness x efficiency). The period of time needed to reach the half unit manufacturing cost by setting annual percentages for MCI is a managerial choice imposed by market conditions (price and profit) and less by the company's internal conditions. This requires the provision of resources such as:

   a. Providing people with the necessary time to participate in *kaizen*, *kaikaku* or problem-solving projects that are continually driven by the need to achieve manufacturing target improvement cost by correctly establishing the workload of people with the operational tasks and tasks to accomplish improvements;

   b. Timely supply of all necessary materials for improvement;

   c. Timely provision of knowledge to maintain the culture of continuous improvement;

   d. Continuous promotion of contextually desirable managerial behavior—Management Branding (MB)—to maintain the creative and innovative state of spirit and people's desire to continually make systemic and systemic changes for the better (Posteucă, 2011; Posteucă and Sakamoto, 2017, pp. 240–244; Posteucă, 2018, pp. 227–230); and last, but not least,

   e. The discipline of all the people within and outside the company, at the level of external collaborators.

Therefore, the continued understanding of the cost reduction opportunities of the main processes of each product family, especially the opportunities for transformation cost and material cost (through design, through alternative materials and suppliers, through successive negotiations with suppliers and by reducing/eliminating unnecessary use of materials) is a continuous challenge for managerial teams.

The basic logic of achieving multiannual manufacturing target profit through productivity is relatively simple: (1) if sales are rising, then increase the number of products manufactured and sold—increase effectiveness, especially for equipment and implicitly maximizing outputs—because the annual average target unit profit is deemed to be most advantageously achieved mainly by improving effectiveness; (2) and if sales are declining, then it is mainly necessary to reduce the manufacturing cost—increase efficiency, especially for human work, material and utilities and, implicitly, minimizing inputs—because the annual average target unit profit is most advantageously achieved by improving efficiency. The cardinal rule of external and internal manufacturing profit is that multiannual manufacturing targets profit do not change irrespective of the evolution of sales (Posteucă and Sakamoto, 2017, pp. 82–83; Posteucă, 2018, pp. 13–14).

Going forward, to ensure the required level of productivity, cost and quality at the level of one unit of product over a year, starting from Figure 1.1 and Formula 1.1, it can be said that:

$$\text{atSPU} - (\text{atEPU} + \text{atIPU}) = \text{VACU} + \text{NVACU} \qquad (1.2)$$

where:

atSPU	= Annual Target Selling Price per Unit
atEPU	= Annual Target External Profit per Unit
atIPU	= Annual Target Internal Profit per Unit
VACU	= Value-Added Costs per Unit
NVACU	= Non-Value-Added Cost per Unit

At the same time, for a family of products or for all products of a manufacturing company, one can say that:

$$(\text{atSPU} - \text{atPU})^* \text{atNUS} = \text{atMCU}^* \text{atNUS} \qquad (1.3)$$

where:

atSPU	= Annual Target Selling Price per Unit
atPU	= Annual Target Profit per Unit
atNUS	= Annual Target Number of Units to be Sold
atMCU	= Annual Target Manufacturing Cost per Unit

In conclusion, the knowledge, awareness, and approach, through systemic and systemic improvements, of the costs related to non-effectiveness or losses (not effectively used input) and the costs related to non-efficiency or waste (excess amount of input) represents the approach in this book of ensuring multiannual manufacturing target profit. In this way, by coordinating improvements through the need and opportunity for MCI, the seemingly antagonistic relations between cost, productivity and quality are defused, and these three elements are simultaneously addressed by MCI to satisfy customers and shareholders through a continuous and consistent approach to CLW improvement.

1.5 THE BASIC CONCEPT OF THE MCPD SYSTEM

From the MCPD system perspective, customer satisfaction at unit cost/unit price, quality (quality control–product and quality assurance–process) and delivery times is addressed by continually targeting all productivity improvements to achieve MCI.

The concept of Manufacturing Cost Policy Deployment (MCPD) is defined as the process of translating the strategic objective of reducing manufacturing costs in the long run toward the improvement of annual systematic activities and toward annual systemic improvement actions by setting targets and means to improve process costs of families of products, in order to

1. Set annual MCI targets and means for all the main processes of all product family cost (PFC);
2. Fulfill the annual manufacturing improvement budgets (AMIB–for both existing products and new products) by continuous investigation of the relationships between costs, processes, and losses and waste;
3. Achieve performance of annual manufacturing cash improvement budget (AMCIB);
4. Direct and plan the systematic and systemic annual manufacturing improvements through continuous reconciliation between the need to reduce costs and process-level opportunities for MCI;
5. Engage the workforce to meet annual MCI targets and means;
6. Measure and analyze performance for MCI;
7. Achieve cost targets at shop floor level (Posteucă, 2015, p. 65; Posteucă and Sakamoto, 2017, p. 81–82; Posteucă, 2018, p. 10–11).

MCPD focuses on CLW for uncovering hidden reserves of manufacturing profitability by improving productivity wherever it is more feasible at process level to ensure consistent product competitiveness through price and profit. Together with these internal benefits—increasing profit by continuous improvement of productivity and consistent product competitiveness through price—there is an external benefit in terms of reducing the environmental impact of unnecessary resource use (material, utilities, forced lifecycle reduction of equipment) to achieve products according to the "voice of customers" and to achieve sustainable development of manufacturing companies.

Within the MCPD system, all the losses and waste of the main processes of each product family and their associated costs are tracked at the level of:

- *Time-related loss (TRL)*: costs of equipment losses and costs of human work losses; and
- *Physical loss (PL)*:

- Costs of technological and material scrap losses;
- Cost of auxiliary consumables losses;
- Costs of energy losses;
- Costs of die, jig, and tool losses;
- Costs of finished products stock, waste (exceeding the number of stock days standard for finished products);
- Costs of raw material stock, waste (exceeding the number of stock days standard for raw material); and
- Costs of components stock, waste (exceeding the number of stock days standard for components) (Posteucă and Sakamoto, 2017, p. 116–127; Posteucă, 2018, pp. 83–85).

The MCPD system by identifying and continuously improving MCI (with the help of CLW and especially CCLW–the root cause of CLW along the manufacturing flow) identifies added-value costs for customers relating to products sold, and free added-value costs associated with losses and waste. Many manufacturing companies are not fully aware of the magnitude of CLW because they do not have a system for capturing CLW data and information and a system targeting improvements in manufacturing profit by improving productivity. These companies still consider that the percentage of CLW in the cost of the product is "acceptable" compared to the current profit, the satisfying profit level and that it is not worth the effort to create a system for identifying and continuously reducing CLW. However, improving long-term productivity involves both increasing the use of current and future capabilities by increasing process effectiveness (inputs already available to manufacturing companies, especially equipment, devices, tools, dies, jigs, losses) as well as reducing the need to acquire new resources by increasing efficiency in and especially between processes (inputs that are scheduled to be achieved, such as human work, raw materials, utilities, components, packaging materials; losses and waste; waste with the seven main types–transport, inventory, motion, waiting, over-processing, overproduction, defects; and waste found especially in standard inventory overflows).

Often, full awareness of the current and potential level of CLW encourages top management teams to allocate all necessary resources (people's time for their participation in the establishment, implementation, and monitoring of improvements and materials needed to achieve improvements) for the successful achievement and implementation of systematic (*kaizen*) and systemic (*kaikaku*) improvements to achieve manufacturing target profit and/or eliminate environmental impacts of current and/or future activities.

By continually quantifying the *losses in costs*, i.e. the unnecessary labor hours of people's work, the minutes of unnecessary operation of equipment, the percentage of raw materials, auxiliary materials, tools and utilities employed unnecessarily, and of *waste (stocks/inventory) in costs*, and stocks that are too expensive and too high at a given time respectively, *annual and especially multiannual profitability scenarios are developed concurrently with associated productivity scenarios* (quality improvement scenarios are included in productivity scenarios). In this context, the profitability and productivity of manufacturing companies in the short-, medium- and, especially, long term, is due to the knowledge and continuous improvement of the CLW through the continuous improvement of productivity (effectiveness and efficiency), coordinated by the current and the necessary level of manufacturing costs, and by establishing manufacturing target profits at the same time as productivity targets. In this way, seemingly antagonistic relations between cost, quality, and time imposed by customers at product level and, implicitly, at product process level are simultaneously addressed to ensure the necessary competitiveness of the company in the long run. By defusing these relationships, a consistent level of medium- to long-term profitability is ensured (see Figure 1.7).

The acceptable level of cost targets, especially manufacturing costs, for product quality and manufacturing times, and delivery times in

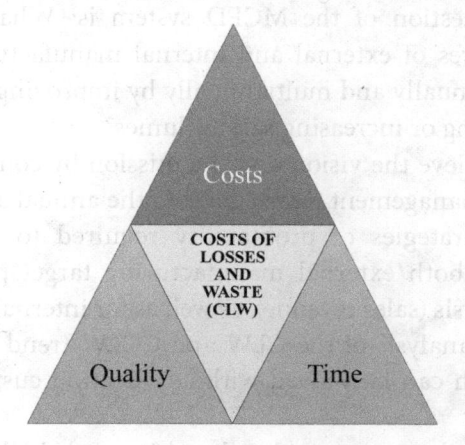

FIGURE 1.7
Reducing and/or eliminating antagonistic relationships between cost, quality and time by setting and meeting multiannual and annual targets for CLW.

particular become the core ingredients of manufacturing target profit and productivity scenarios and strategies.

The MCPD system is applicable to all manufacturing companies, regardless of the type of industry (*process industry*–food, textiles, oil, rubber, glass, paper and pulp, chemicals, etc. or *manufacturing and assembly industry*–automotive, machinery, precision instruments, electrical products, metal products, aerospace, precision instruments, etc.), which performs the transformation of inputs into outputs mainly with the help of the equipment, but also with the help of operators and using materials, tools and energy utilities of any kind and in any quantities along the manufacturing flow. Even though the MCPD system is designed for a manufacturing company, its approach can be extended to multiple companies throughout the supply chain, companies that provide external logistics to achieve an integrated CLW approach and, implicitly, total operating profit.

In this context, the setting of annual targets and means pertaining to CLW at the level of each PFC process in conjunction with targets and means for OMIs and other KPIs in the company alongside CLW-related KPIs represents a challenge that addresses the entire MCPD system.

Therefore, as one can see in Figure 1.8, the MCPD system is a three phase and seven step approach of CLW and CCLW at the level of the main processes of each PFC to establish and meet annual MCI targets and means to achieve consistent annual and multiannual manufacturing target profit as follows (the three phases and the seven steps will be discussed in more detail in Chapter 2) (Posteucă, 2018, p. 57–60):

The central question of the MCPD system is: What are the profit contribution shares of external and internal manufacturing profit that can be earned annually and multiannually by improving productivity in terms of decreasing or increasing sales volumes?

In order to achieve the vision and the mission by commitment to the MCPD system, management teams develop the annual and multiannual scenarios and strategies of profitability required to be obtained by productivity for both external manufacturing target profits amid the sales trends analysis (sales revenue), as well as for internal manufacturing profit, amid the analysis of the CLW and CCLW trend (manufacturing costs, costs which can be waived without affecting customer perceived value).

Going forward, the core annual and multiannual challenge of manager teams is to set targets and means for CLW and CCLW to consistently perform MCI for each PFC amid the compliance with the core principles

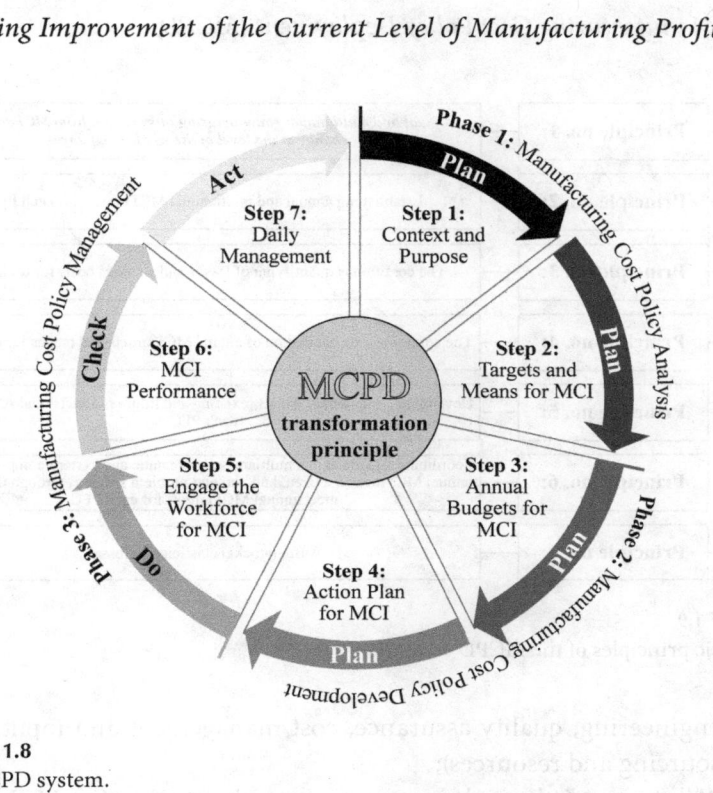

FIGURE 1.8
The MCPD system.

of the MCPD system (Posteucă and Sakamoto, 2017, pp. 82–85; Posteucă, 2018, pp. 13–15).

In Figure 1.9 the seven basic principles of the MCPD system are presented.

Principle 1 is the cardinal principle of the MCPD system.

The MCPD system slogan that best captures its intentions is: "Everyone, everywhere, all the time, together, in the same direction, and all the same." Over time, this slogan supports the creation and continued development of the pro-productivity identity of managers and all those involved in the MCPD system to continuously achieve company's productivity vision through the pro-MCPD culture. Slogan explanations are as follows:

- *"Everyone"*: the total involvement of all people in and outside the company to meet MCI targets and means;
- *"Everywhere"*: the CLW and CCLW approach of all the main processes of each PFC, both at the shop floor level (manufacturing and maintenance) and manufacturing process level (inputs from development and design, manufacturing control, manufacturing

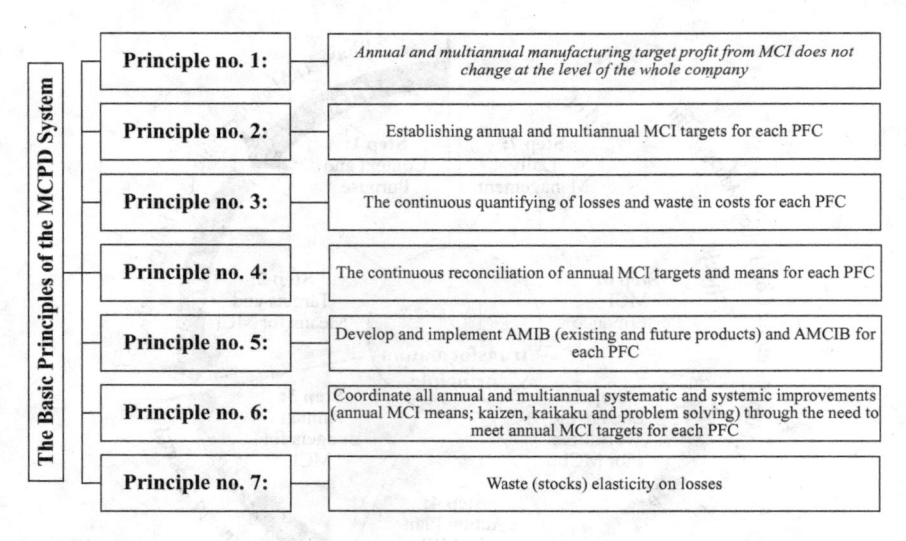

FIGURE 1.9
The basic principles of the MCPD system.

engineering, quality assurance, cost management and inputs from sourcing and resources);

- *"All the time"*: the real ongoing concern about achieving MCI targets (there must be continuous systematic and systemic improvement projects in preparation, ongoing monitoring, and evaluation of the results of real reduction of manufacturing unit costs and consistent profitability in the medium and long term; each employee participates in at least one strategic improvement project per year and at least 10 improvement ideas within daily manufacturing cost improvement management);
- *"Together"*: refers to the fact that a systemic and/or systematic improvement aimed at achieving MCI targets is not achieved by an individual or by a department, but always by an interdisciplinary team that provides clear and objective directions for achieving MCI targets;
- *"In the same direction"*: refers to the fact that all *kaizen* and *kaikaku* improvements are converging with the need to achieve annual and multiannual MCI targets for each PFC and, implicitly, the manufacturing profit and the manufacturing unit cost reduction;
- *"All the same"*: it is about creating a desirable contextual behavioral identity at the level of managers and at the level of each employee in and outside the company to use the same methods, techniques and tools to achieve the exact tangible and intangible effects needed and, especially, the annual and multiannual MCI targets for each PFC.

1.6 CONCLUSIONS

In conclusion, MCPD is not a costing method, but an MCI system. The MCPD system's goal for manufacturing companies is to achieve annual and, especially, multiannual manufacturing profits consistently under productivity controlling and targeting in the medium-term and long-term with the help of an interdepartmental organization coordinated by the need to meet MCI targets with the background of a culture and work environment that supports maximizing outputs (effectiveness) and minimizing inputs (efficiency) of all the core processes of each PFC.

The development and implementation of annual and multiannual profitability scenarios by improving productivity, amidst the potential increases or decreases in sales volumes of each PFC, contributes to the harmonious transformation of the manufacturing flow and to the MCPD's main stake in securing the external, but especially the internal target profit, by conducting systematic improvement profitable projects (*kaizen*) and systemic improvement projects (*kaikaku*) converging continuously to the need to achieve CLW and, especially, CCLW. In this way, a concurrent approach to the need to reduce costs, and improve quality and time is achieved by continually developing and updating the annual and multiannual productivity master plan.

The more unstable and possibly decreasing the equipment's capacity Overall Equipment Effectiveness or OEE is, the higher the waste level (associated to all inventory categories), the level of losses (associated to human work, material, utilities), and associated cost of losses and waste (CLW) level, which means higher product unit cost of manufacturing and lower annual and multiannual profit.

Reducing CLW is the basic need of manufacturing companies.

2

Scenarios and Strategies of the MCPD System

Establishing MCI targets and targeting all systematic (*kaizen*) and systemic (*kaikaku*) improvements by coordinating them according to the need to meet MCI targets requires establishing the framework within which the manufacturing companies will operate. Specifically, it requires setting down scenarios and strategies for the subsequent development of the manufacturing system according to market signals. By developing scenarios to achieve MCI targets and means, it is sought to describe plausible future alternatives to address credible and consistent CLW and CCLW, depending on the potential sales situation and, consequently, depending on the necessary pressure on domestic and external profit to achieve CPV (annual and multiannual manufacturing target profits and target sales volumes). By developing and implementing strategies to achieve MCI targets and means, it is sought to identify concrete ways of implementing the scenario chosen for the future to meet MCI targets and means by effectively and efficiently planning and mobilizing all the necessary resources both within and outside the company by developing and implementing PMP.

Therefore, in the first section of this chapter, Section 2.1, we present the connections between the external manufacturing target profits through maximizing outputs and the internal manufacturing target profits through minimizing inputs with four potential scenarios of sales volume evolution to create the prerequisites for the manner of establishing business expectations and productivity strategies to achieve MCI targets for each PFC, depending on the current and future business context. Section 2.2 presents MCI change drivers for each PFC—these are the elements through which successive or systemic improvements can generate significant beneficial changes over the manufacturing system and MCI in the four sales scenarios for both external manufacturing profits and internal manufacturing profits. It also presents system elements for each MCI change driver of each important process of each PFC to identify the elements and areas where close scrutiny is required to continuously identify and measure losses and waste and associated CLWs in order to establish basic MCI strategies. Section 2.3 outlines the necessity and direction of continuous survey of each important process of each PFC to determine the total current level of CLW for each system element of MCI change drivers and presents the method of developing MCI annual and multiannual strategies by addressing the current level of CLW for each PFC process in order to further establish MCI targets and means. Section 2.4 identifies current assumptions for CCLW and defines the most credible assumptions for MCI strategies. Section 2.4 describes the method to approach MCI targets and means based on current assumptions for previously identified CCLWs. Section 2.5 presents tangible and intangible effects of the MCPD system based on the development of scenarios and structured strategies to achieve MCI targets by continuously and harmoniously converting the manufacturing flow of each PFC. Finally, Section 2.6 presents the conclusions reached in this second chapter.

2.1 PROFITABILITY SCENARIOS HYPOSTASES OF THE MCPD SYSTEM IN CASE OF INCREASE OR DECREASE OF SALES

From the perspective of the MCPD system, as mentioned in the previous chapter, annual and multiannual manufacturing target profit scenarios represent alternatives to addressing the achievement of the annual MCI targets and means by continuously improving the productivity of the processes of each PFC, depending on the increasing trend in sales and the predominant need to maximize outputs (E), or the downward trend in sales and the preponderant need to minimize inputs (I).

The basic logic of the scenarios for MCI is: product unit cost of manufacturing tends to decrease proportionally with the decrease of losses and waste from process and, implicitly, from CLW. The more CLWs and better CLWs are identified and addressed across the main processes of each PFC, the more likely they are to set and achieve the annual and multiannual MCI targets through MCI means.

Figure 2.1 presents the structured development of scenarios through the MCPD system to provide the framework for setting MCI targets and means for the main processes of each PFC.

	Elements	Scenario 1: major and flexible sales growth	Scenario 2: incremental sales growth	Scenario 3: incremental sales declines	Scenario 4: major unexpected sales declines
1) Scenarios Hypostases	External Manufacturing Profit through Maximizing Outputs (E)				
	Internal Manufacturing Profit through Minimizing Inputs (I)				
2) Change Drivers of MCI	CD1 - Effectiveness of current equipment (E)				
	CD2 - Effectiveness of new equipment (E)				
	CD3 - Development of new profitable products (E)				
	CD4 - Maximizing variable cost efficiency (I)				
	CD5- Maximizing fixed cost efficiency (I)				
	CD6- Continuously improving manufacturing lead time (I)				
	CD7- Continuously aligning processes to market needs (I)				
	CD8- Continually improving inventory levels (I)				
3) System Elements of Change Drivers of MCI for each PFC	System element 1 for D1: Equipment Availability				
	System element 2 for D1: Equipment Performance				
	System element 3 for D1: Equipment Quality				
	System element 4 for D1: Equipment Shutdown				
	System elements for D2-D7:....				
4) Current level of CLW for each System Element of Change Drivers of MCI	Cost of Equipment Losses (%, hours and $)				
	Cost of Human Work Losses (%, hours and $)				
	Cost of Material Losses (% and $)				
	Cost of Energy/Utilities Losses (% and $)				
	Cost of New Products Development Losses (% and $)				
	Cost of New Equipments Development Losses (% and $)				
	Cost of Waste: WIP from Set-up (WIP S) (minutes) or units				
	Cost of Waste: WIP from Transfer (WIP T) (minutes) or units				
	Cost of Waste: Raw Material Inventory (minutes) or units				
	Cost of Waste: Finished Products Inventory (minutes) or units				
	Cost of Waste: Componentes Inventory (minutes) or units				
	Cost of Waste: Packaging Inventory (minutes) or units				
5) The basic MCI Strategies for each current level of CLW	Strategies for opportunities of the CLW 1 of CD1				
	Strategies for challenges of the CLW of CD1				
	Strategies for opportunities of the CLW of CD2				
	Strategies for challenges of the CLW of CD2				
	Strategies for opportunities of the CLW of CD3....CD8				
	Strategies for challenges of the CLW of CD3....CD8				
6) Current assumptions to identify CCLW	Credible assumptions for each MCI strategy (CD1...CD8)				
	Uncertain assumptions for each MCI strategy (CD1...CD8)				
	Vulnerable assumptions for each MCI strategy (CD1...CD8)				
7) MCI Targets	MCI targets for opportunities of the CCLW of CD1-CD8				
	MCI targets for challenges of the CCLW of CD1-CD8				
8) MCI Means (kaizen & kaikaku)	MCI means for opportunities of the CCLW of CD1-CD8				
	MCI means for challenges of the CCLW of CD1-CD8				

FIGURE 2.1

Development of structured scenarios to achieve the continuous transformation of the manufacturing flow of each PFC through the MCPD system.

As one can see, in section 1 of Figure 2.1, there are the two profitability scenarios hypostases (boundaries for the scenario) in connection with the four cases of sales scenarios. The first two sales scenarios (Scenario 1: major and flexible sales growth; Scenario 2: incremental sales growth) aim to focus on multiannual and annual manufacturing target profits by concentrating especially on external manufacturing profits through maximizing outputs (E). The following two sales scenarios (Scenario 3: incremental sales declines; Scenario 4: major unexpected sales declines) aim to focus on achieving multiannual and annual manufacturing target profits by focusing on internal manufacturing profits through minimizing inputs (I).

Therefore, for the development of structured scenarios to achieve the continuous transformation of the manufacturing flow for each PFC through the MCPD system, the starting points are the two hypotheses of the need to transform the manufacturing flow by improving productivity (E and I: boundaries for the scenario; section 1 in Figure 2.1) by detailing the *change drivers* (D1–D3 for "E" and D4–D8 for "I"; change drivers describe how they contribute to E and I and the way they can be modeled according to the logic of the "E" or "I" scenario—section 2 of Figure 2.1), then define the *system elements* for each change driver individually (or the manner of expression of each change driver—section 3 in Figure 2.1), then the current CLW for each element system is established (section 4 in Figure 2.1), then the *basic MCI strategies* are set for each opportunity and challenge of CLW (which may occur for each system element of each change drivers—section 5 in Figure 2.1) so that the *annual and multiannual MCI targets and means* (*kaizen* and *kaikaku*—sections 7 and 8 in Figure 2.1) are finally established on the basis of the current assumptions underlying the establishing of CCLW (section 6 of Figure 2.1). Interdepartmental MCPD teams will analyze and test the current assumptions for CCLW to verify the robustness of assumptions for the selected sales scenario for the following period (of the four possible sales scenarios) for three types of assumptions: (1) credible or acceptable for planning the future MCI targets and means based on current CCLW; (2) uncertain—they require further analysis to establish the MCI targets and means based on current CCLW; and (3) vulnerable—they should not be taken into account when planning the framework for MCI targets and means based on current CCLW.

Figure 2.2 shows the four possible scenarios of sales.

Note: Scenarios 1 and 2 are desirable. Scenario 1 involves setting and achieving MCI targets imposed by the market, with increased attention

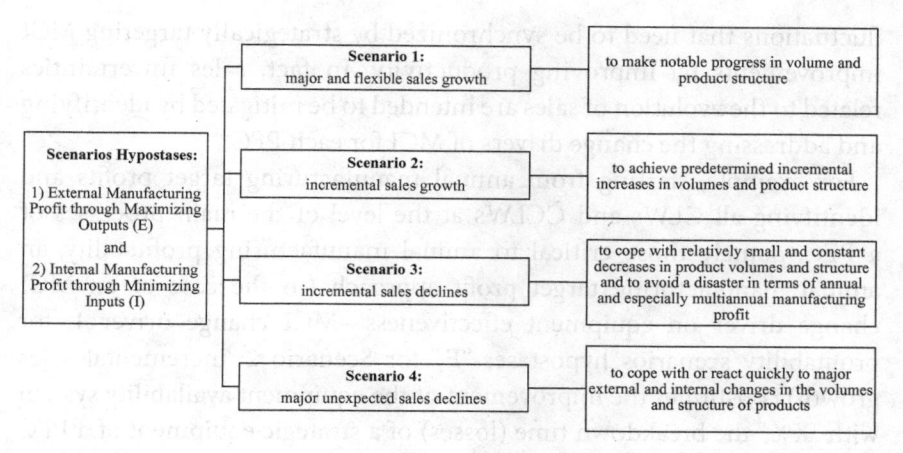

FIGURE 2.2
Profitability scenarios hypostases and the four sales scenarios of the MCPD system.

to effectiveness and method design (products, equipment and processes) (Posteucă and Sakamoto, 2017) to achieve *the right things* (Drucker, 1963), but also an increased focus on efficiency to achieve *the right things* by developing MDC improvement projects (Posteucă and Sakamoto, 2017, p. 327–354), *kaikaku* and *kaizen*. Scenario 2 involves mastering internal and external elements of the company, consistent planning over time, and implementation of predominantly *kaizen*-type improvement projects.

Each manufacturing company determines what each of the four scenarios means for it. For example, for a certain company, an increase of 15% per year in sales may be considered incremental growth (Scenario 2), while for another company this 15% increase may be a major increase in sales (Scenario 1) and one which implies a significant increase in flexibility and manufacturing flow capacities. These four sales scenarios aim at defining key assumptions and explaining the hidden risks of not meeting both external and internal manufacturing target profits, both at the level of the whole PFC and at the level of each PFC process (Posteucă, 2018, pp. 117–121).

Over a year or a period longer than one year, any company must define a single probable sales scenario and then the method to boost productivity to achieve multiannual and annual manufacturing target profits through multiannual and annual external and internal manufacturing target profit (Posteucă and Sakamoto, 2017, pp. 67–79). This probable sales scenario depends on the planned lifecycle validity of the product, equipment, and technology and seeks to identify the types of external and internal fluctuations that need to be synchronized by strategically targeting MCI improvements by improving productivity. In fact, sales uncertainties related to the evolution of sales are intended to be mitigated by identifying and addressing the change drivers of MCI for each PFC.

For example, starting from annual manufacturing target profits and identifying all CLWs and CCLWs at the level of the main processes of a PFC, namely those critical to annual manufacturing profitability, an annual manufacturing target profit approach (in the case of an MCI change driver on equipment effectiveness—MCI change driver 1, for profitability scenarios hypostases "E," for Scenario 2: incremental sales growth) constitutes the improvement of the equipment availability system with "X%," the breakdown time (losses) of a strategic equipment of a PFC over the next 12 months by reducing the cost of equipment losses by "Y%," which is: (1) a reduction by "Z%" of the CCLW summed up across all the main PFC processes; (2) a reduction by "W%" of product unit cost of manufacturing; and (3) an increase by "T%" of annual manufacturing target profit (taking into account the types of products planned to be achieved).

System elements are the elements that have the greatest potential for change or interruption during the analyzed time period (at least 12 months) associated with the impact of the change drivers. Examples of system elements for breakdown time (losses) of a PFC strategic equipment can be: the number of breakdown events, the average duration of the breakdown events, the cost of the spare parts, the MTTR-related cost, the opportunity cost of the unrealized production, the cost of overtime, the cost of direct and indirect labor affected by equipment breakdown, the WIP inventory cost above the predetermined standard level, etc.

Therefore, with the help of the development of scenarios hypostases of the MCPD system in one of the four sales scenarios, the following can be achieved:

- Strengthening interdepartmental/interdisciplinary skills to identify the main challenges and opportunities and to improve and innovate the manufacturing flow of a PFC on a continuous basis by observing and testing the future alternatives of the main processes to achieve annual and multiannual MCI targets;
- Ensuring a deep understanding of the manufacturing system and the market by analyzing the main trends and the most credible working hypotheses;

- Developing the framework for the objective setting of credible assumptions for each MCI strategy based on the analysis of the most likely future important market changes and the manufacturing system;
- Developing the framework for establishing annual and multiannual MCI targets and means (rigorous, systematic, systemic, verifiable and based on an analysis of all the possibilities of action known at that time);
- Attracting people to share their knowledge visually at every step of the MCPD scenario development process to continuously support future MCI targets;
- Continuous development of a clear sense of the challenges and urgent opportunities among all the people in the company and the rapid support of operational, tactical and strategic decisions;
- Increasing the objectivity of MCPD scenarios by continually rebuilding them through working in interdepartmental groups.

The final aim of the scenarios is to transpose them at the level of each key process of each PFC in order to establish the annual and multiannual framework of MCI targets and means. Each PFC and related main process has an annual scenario described in words based on structural scenarios of one paragraph or half a page. This narrative description helps achieve the objective of inter-departmental awareness of opportunities and challenges in achieving the annual and multiannual MCI targets.

2.2 DEVELOPED MCI CHANGE DRIVERS, SYSTEM ELEMENTS AND BASIC MCI STRATEGIES TO SUPPORT EXTERNAL AND INTERNAL MANUFACTURING PROFIT

The two profitability scenarios hypostases of the MCPD System ("E" and "I" in Figure 2.1) and the four forms of sales scenarios constitute a method of visualizing how manufacturing companies evolve over time and how MCI change drivers interact with each other. MCI change driven are the mechanisms in the manufacturing system that influence "E" and "I" in the four sales scenarios.

Figure 2.3 presents MCI change driver for profitability scenarios hypostases of the MCPD system.

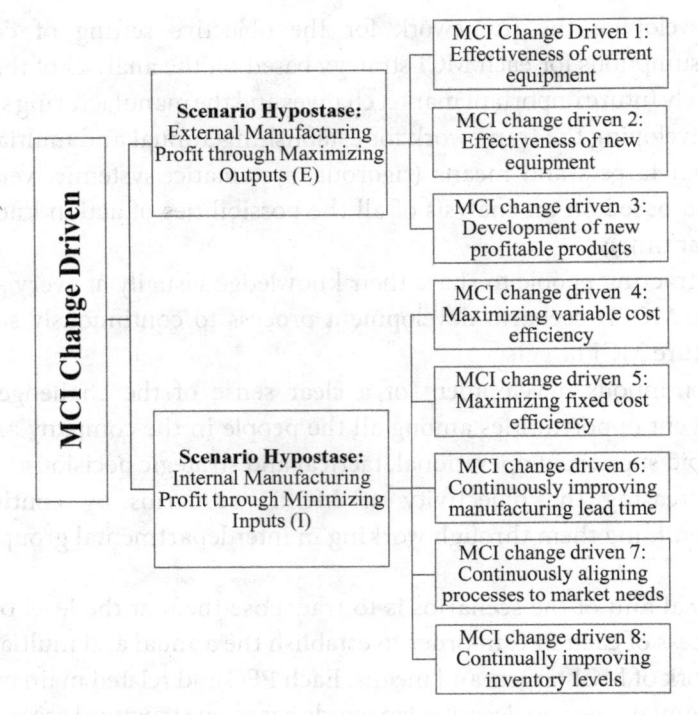

FIGURE 2.3
MCI change driver for profitability scenarios hypostases of the MCPD system.

MCI change drivers are the factors with the greatest impact on "E" and "I" and which require the development of individual system elements depending on the likely evolution of sales. MCI change drivers describe the manner in which they contribute to the achievement of "E" and "I" and the method by which they can be shaped according to the logic associated with profitability scenarios hypostases (predominantly "E" or "I").

The MCI change drivers, system elements and basic MCI strategies for each PFC for the two major scenarios hypostases related to the likely trend of sales, growth or reduction are presented below. These are scenarios that require different approaches to productivity improvement to achieve the annual and multiannual manufacturing targets profit. The scenario approach at the level of each PFC process with a view to establishing in detail the MCI targets and means will be presented in Chapter 3 and examples will be given in Chapter 4. The present chapter presents the definition of the framework for establishing MCI targets and means through scenarios and strategies for MCI.

2.2.1 External Manufacturing Target Profit through Maximizing Outputs

Therefore, to achieve the annual manufacturing target profit for each PFC by focusing in particular on the external sales profits, against the backdrop of an increasing trend in current and future sales, it is necessary to develop MCI change drivers of MCPD converging with CPV and CPM by paying special attention to maximizing outputs, especially equipment that provides the volume of products needed in manufacturing companies, and attention to maintaining at least at a constant level the inputs from processes.

Concentration of productivity improvement in the maximizing outputs scenario is *focused on losses* or not effectively used input, but also requires a focus on minimizing inputs or efficiency improvements or, in other words, limiting or reducing the excess amount of input or waste.

From the perspective of maximizing annual outputs, MCI change drivers aim to maximize effectiveness by paying special attention to equipment that delivers product volume and the continuous development of new, profitable products on time.

Savings that can be made on the cost of losses associated with equipment and new products do not have a direct impact on the profit and loss account but can be budgeted at AMIB and AMCIB on the basis of annual MCI planned and implemented (*kaizen* and *kaikaku*).

In this regard, from the perspective of effectiveness, starting from Peter F. Drucker's concept of "doing the right things" (Drucker, 1963), it can be extended to:

1. The right product;
2. The right equipment;
3. The right availability (of the equipment);
4. The right time (of the equipment);
5. The right quality (of the equipment), and
6. The right place (in the process).

Developing the maximizing outputs scenario ("E") to achieve the annual and multiannual MCI targets and means involves defining change drivers and associated key assumptions, and explaining the hidden risks of achieving the *product number increase* by detailing expectations related to:

1. Effectiveness of current and new equipment, and
2. The development of new profitable products.

Therefore, the right product and the right equipment or product and equipment effectiveness are essential to ensure productivity levels and, implicitly, external manufacturing targets through maximizing outputs. As Peter F. Drucker noted with regard to *doing the right things and doing things right* (Drucker, 1963), doing the right things is more important than increasing efficiency in the wrong way—inappropriate products and equipment continuously generate unnecessary losses and waste along the entire manufacturing flow; this is implicitly a high level of product unit cost of manufacturing. Therefore, by increasing efficiency through *kaizen* projects, a lower product unit cost of manufacturing can be achieved, especially by reducing non-value-added activities (waste), but the questions arising are: Why are those activities/costs arising? Have they been well-designed to meet the needs of the customers? Why should non-value-added activities (NVAAs) that might have been eliminated through a solid design occur?

Choosing or designing a piece of equipment and designing a new product are the key elements of product number increase and external manufacturing targets profit through maximizing outputs. Furthermore, for example, choosing and implementing a *kaikaku* project with new equipment to replace old equipment that no longer meets a customer's external requirements and/or internal requirements will ensure the level of effectiveness and efficiency only through programming and implementing *kaizen* improvement projects; this *kaikaku* project for new equipment should ensure a low level of losses and waste through a proper design of the equipment. We will deal with this in further detail in Section 2.2.1.2. Consequently, the importance of design is overwhelming in the case of new products and in all processes (manufacturing methods) to ensure the transformation of inputs into outputs effectively and efficiently (Posteucă and Sakamoto, 2017, pp. 257–376).

2.2.1.1 Maximize Outputs of the Entire Manufacturing System by Continuously Improving the Effectiveness of the Current Equipment

This *MCI change driver* related to increasing effectiveness of the current equipment focuses primarily on external manufacturing target profit through maximizing outputs ("E") and has the following main goals:

- Reducing and/or removing equipment breakdown;
- Reducing setup time;

- Aligning the equipment cycle time to the takt time;
- Continuous improvement of mean time between failure (MTBF) and mean time to repair (MTTR);
- Eliminating handling issues;
- Implementing *poka-yoke* (mistake-proofing) devices to reduce scrap and rework levels;
- Reducing or eliminating overtime; and last but not least,
- Reducing or eliminating the failure to achieve the planned target manufacture.

In fact, maximizing equipment outputs by increasing equipment effectiveness aims at increasing OEE and, implicitly, continuous and systematic (*kaizen*) improvement and targets the following *system elements*: (1) failure losses; (2) setup/adjustment losses; (3) cutting-blade losses; (4) start-up losses; (5) minor stoppage/idling losses; (6) speed loss; (7) defects and rework losses; and (8) shutdown losses.

Therefore, the basic MCI strategies needed to identify the opportunities and challenges of increasing effectiveness of the current equipment are aimed at:

- Establishing the percentage of periodic increase of the current level of OEE throughout the entire manufacturing system for each PFC (for strategic equipment) over a defined period (one year or several years) by reconciling the needs imposed by the market (product volumes, profit and price level) and the opportunities to achieve the annual and multiannual MCI targets, this in fact the level of CLW and CCLW that can be approached through annual and multiannual MCI means, namely through the successful planning and implementation of profitable projects of systematic improvement (*kaizen*) for each type of equipment loss, in order to achieve the increase in the number of target products and the manufacturing profit targets according to CPM and CPV;
- Establishing the best time to switch to new equipment to meet external (customers, suppliers and competitors) and internal (CLW) demands.

Depending on the *basic MCI strategies*, continuously improving the effectiveness of the current equipment coordinated with the need to achieve the annual and multiannual MCI targets that require the transformation of all equipment losses in costs for each PFC to be determined, and afterward the CCLW identification to establish CCLW targets by establishing equipment losses targets, and the necessary systematic (*kaizen*) and systemic (*kaikaku*) improvements (MCI means targets). The ultimate goal of this equipment effectiveness-related change driver is to identify the most credible and feasible assumptions for each MCI strategy to increase the objectivity of establishing the annual and multiannual MCI targets and means for a representative timeframe in the future (on a continuous basis for at least 12 months in advance) in one of the four possible business scenarios presented in Figure 2.1.

2.2.1.2 Maximize Outputs of Future Equipment to Meet the Demands of Current and Future Products

This *MCI change driver* related to increasing the effectiveness of the future equipment primarily focuses on securing the external manufacturing target profit through maximizing outputs (E) and has the following main goals:

- Optimal choice of time to switch to new equipment (purchase or development of new equipment);
- Choosing the capacity of the equipment that will contribute to achieving the target product volumes planned in the next period (according to product life cycles; with an acceptable cycle time and set-up time);
- Accurately planning the increase of the load capacity of the new equipment to the optimum state;
- Choosing the equipment that will have a low level of defects;
- Choosing the equipment that has an acceptable life cycle cost (LCC);
- Choosing the equipment to continuously support innovative and integrated technologies and products in the current flow of manufacturing technology;
- Choosing the equipment to continuously support environmental management.

The main system elements related to a new equipment (for 2–3 possible options for choosing the new equipment) are:

- The need for the new equipment in terms of capacity demanded over time (increasing OEE over time, taking into account the number of working days per year, the number of shifts, the normal life of the equipment, the demand for products according to the life cycle of the products which will be created with the help of the equipment, the probable level of the scrap, and last but not least, the standard cycle time and standard set-up time);

- The time of delivery of the new equipment (depending on: equipment created within the company, standard equipment from the supplier or equipment from the supplier with design changes);
- The time of installation of the new equipment and of the production of the first pieces declared to be of the quality required;
- The cost of the equipment (purchase cost, transportation cost, installation cost and technological samples);
- The cost of operating the equipment (costs of: raw materials and consumables, direct labor, maintenance, energy, spare parts, aiding assistive devices, staff training, dismantling of the equipment, residual value planned, environmental impact, e.g. pollution, etc.).

The two or three possible variants of *the main system elements* of a new equipment must provide two to three LCC alternatives for the average unit cost of the products to be made with that equipment. Choose the equipment that will offer the lowest product unit cost of manufacturing consistently over time.

In this context, *basic MCI strategies* for new equipment to identify the opportunities and challenges of increasing effectiveness will target:

- Capacity needs (congruence between the capacity of the equipment and the necessary capacities of the manufacturing system in the future in accordance with CPV and CPM);
- Delivery date;
- Initial and operating cost (LCC);
- The required level of flexibility (set-up time);
- Operators' safety in the operation of the equipment;
- The possibilities of reducing the cycle time of operators by reducing the operator's non-value-added activities (NVAA);
- Dimensions of the equipment to fit in the available layout (event analysis of the cost of possible layout changes);
- The tolerance level of the equipment parameters;
- The compatibility of the raw material with the equipment;
- The complexity of autonomous maintenance (cleaning, lubrication and inspection);

- The complexity of planned maintenance activities (planned interventions);
- The availability and supply lead time for spare parts.

Establishing the best moment to switch to new equipment, in fact from systematic improvement (*kaizen*) to systemic improvement (*kaikaku*) is determined by:

- Anticipating the moment when the capacity of the current equipment with an already high level of OEE (for example, over 84%) will no longer meet the future demands of the customers;
- The current capacity of the equipment is too high compared to the current and future volumes demanded by customers;
- Improving OEE is not feasible, especially amid future increased volume of customers;
- Measuring and improving OEE is uncertain;
- The parameters of the equipment can no longer meet the characteristics of the new products;
- The product quality level can no longer be ensured by the current equipment;
- Equipment maintenance tends to become too expensive (the cost of spare parts) and insecure (the equipment supplier no longer manufactures that equipment, its stock of spare parts is decreasing and/or it no longer manufactures those spare parts);
- Raw materials from current suppliers can no longer be processed with the current equipment;
- MTTR is increasing and the opportunity to achieve the planned manufacturing is lost (CCLW for planned or unplanned maintenance interventions is increasing);
- The consumption of utilities is increasing;
- And last but not least, the cycle time of a piece of equipment is variable and/or can no longer be aligned to the upstream and downstream equipment cycle times—possibly an assembly line—to reach a takt time with an acceptable level of work in process (WIP) (Posteucă and Sakamoto, 2017, pp. 214–224).

Depending on the *basic MCI strategies* of the new equipment and the acquisition or development decision of the new equipment, the most credible and feasible assumptions for each MCI strategy are determined, and *kaikaku*-type projects for changing/installing the new equipment and *kaizen* projects for raising the OEE level of new equipment to handle CPV and CPM in one of the four possible business scenarios shown in Figure 1.9 are arranged. The main types of losses and the associated cost of losses over the entire manufacturing flow of a PFC are: *equipment development/installation delay losses* (TTSM) (failure to meet time-to-start manufacturing or TtSM; the effectiveness of installing the new equipment) and *development inefficiency of the equipment losses* (acceptance of the installation of new equipment that does not meet all of the original requirements and which subsequently causes losses and waste along the entire manufacturing flow on a continuous basis; the efficacy of installing the new equipment).

2.2.1.3 Maximizing the Load of Current and Future Equipment Through the Continuous Development of New Profitable Products

This *MCI change driver* for increasing the effectiveness of the new products primarily focuses on ensuring external manufacturing target profit through maximizing outputs ("E") and has the following main goals:

- Ensuring competitiveness by continuously launching new profitable products for market segments/niches; with the features and characteristics of future products indicated by the marketing department to support sales (CPV and CPM implicitly);
- The validity over time of the life cycle, volumes and planned profit levels;
- Continuous time-to-market (TtM) reduction to increase responsiveness to competitors' moves;
- Material cost reduction through innovative design and design review;
- Reducing the investment costs needed to launch, create and sell the new product;
- Reducing/eliminating CLW from current processes by designing products to improve the effectiveness of current and/or future processes (Posteucă and Sakamoto, 2017, pp. 170–188; Posteucă, 2018, pp. 195–203);
- Increasing man-hours productivity;
- Increasing the load capacity of current and future equipment;
- Compliance with environmental requirements.

The main system elements related to a new product in the logic of MCPD scenarios (items with a high potential for change or interruption) are:

- Time-to-market (TtM);
- Material cost;
- CLW from the current and future processes in which the future product will be achieved;
- Investment costs.

In this context, *basic MCI strategies* for new equipment will target:

- Correct and complete definition of the new product (Ichida, 1996);
- Material cost reduction ratio;
- Increasing the use of raw materials and hybrid components (standardized parts and subassemblies for many products) in the R&D phase (raw materials and components used in several company products, lowering costs through negotiating power with suppliers);
- Time-to-market reduction (or R&D lead time reduction);
- CLW reduction by designing products to support the increasing efficiency and effectiveness of current and future processes;
- Choosing the potential new equipment to create new products to support increasing the efficiency and effectiveness of current and future processes;
- The design of new products that meet environmental requirements;
- Falling into the future target level of man-hours per product;
- First product design (including design for parts and subassembly) and aligning the cost of the new product with unit target costs imposed by both the market and the need for long-term profitability;
- Prototype execution and aligning the cost of the new product with unit target costs;
- First and second trial manufacturing and analysis.

Depending on the *basic MCI strategies* of the new product and the decision to approve the new product, the plan for achieving the multiannual MCI targets is defined (design and processes: reducing the *cost of losses due to the lack of efficiency of the new product development*—more exactly, cost overruns from the originally planned design costs; and reducing *the cost of losses due to the ineffectiveness of the new product*—more exactly, non-fulfillment of the necessary features and functions of the new product and/

or full non-synchronization of the characteristics of the product with the performance of the process parameters of the equipment; the generation of systemic losses and waste) and future *kaizen* projects are planned to reduce CLW of new products from future processes to meet CPV and CPM in one of the four possible business scenarios presented in Figure 2.1.

2.2.2 Internal Manufacturing Target Profit through Minimizing Inputs

To achieve the annual manufacturing target profit for each PFC by focusing on internal profit, especially against the backdrop of a downward trend in current and future sales, it is necessary to develop MCPD converging scenarios with CPV and CPM by paying particular attention to minimizing inputs–efficiency improvement. Concentration of productivity improvement in minimizing inputs is focused on *losses* or not effectively used input for human work, material and energy and *waste* or reducing or eliminating the excess amount of input. Focusing on minimizing inputs does not exclude paying attention to obtaining external profit by maximizing outputs or effectiveness improvement. In fact, in a company there may be more PFCs and each PFC has its own sales trend. Furthermore, in PFC processes, there may be different needs for increasing outputs or reducing inputs.

From the perspective of the annual minimizing inputs, MCI change drivers aim to maximize efficiency, especially through:

1. Maximizing variable cost efficiency;
2. Maximizing fixed cost efficiency;
3. Continuous improvement of the manufacturing lead time and takt time;
4. Continually improving inventory levels: raw materials, components, consumables and finished products (see Figure 2.3).

The CLW associated with minimizing inputs that can be saved have a direct impact on the annual profit and loss account and are taken into account when preparing the AMIB and AMCIB (for example, improving material and utility costs, improving labor costs/working hours, improving maintenance costs, improving non-quality costs, improving the inventory of raw materials and finished products, etc.).

In this sense, from an efficiency perspective, starting Peter F. Drucker's concept of "doing things right" (Drucker, 1963), it can be extended to:

1. Doing things right: variable cost;
2. Doing things right: fixed cost;
3. Doing things right: manufacturing lead time (WIP) and takt time; and
4. Doing things right: inventory levels.

Therefore, the main expectations of internal manufacturing profit through minimizing inputs are related:

1. Reduction of product unit cost of manufacturing;
2. Reduction of manufacturing delivery time (in particular manufacturing lead time);
3. Increasing the level of alignment of each primary process of each PFC at the takt time; and
4. Reducing the inventory level beyond the manufacturing area, but without obstructing product manufacturing due to lack of inventory at the required time.

In order to achieve these goals, a holistic approach to scenarios is needed to reduce manufacturing lead time and manufacturing costs (Posteucă and Zapciu, 2015c).

2.2.2.1 Minimize Inputs by Maximizing Variable Cost Efficiency

This *MCI change driver* related to the increase of the efficiency of the variable costs, especially indirect variable costs, aims to ensure internal manufacturing targets profit through minimizing inputs (I) and has the following main purposes:

- The exact location of variable costs for each main process of each PFC;
- The exact location of the variable costs at the level of each main process of each PFC;
- CLW analysis of each PFC from the variable cost perspective;
- Continuous variable cost reduction for each PFC;

- Increasing the productivity of all the equipment in particular;
- Reducing the set-up time.

The main system elements related to the increase of the possible variable cost are:

- Direct material costs;
- Cost of direct material losses;
- Indirect material costs (auxiliary materials cost for manufacturing);
- Cost of indirect material losses;
- Utility costs (manufacturing overheads; from the transformation cost structure);
- Cost of utility losses;
- Tool costs (manufacturing overheads; from the transformation cost structure);
- Cost of tools losses;
- Die and jig costs (manufacturing overheads; from the transformation cost structure);
- Cost of die and jig losses.

In this context, *basic MCI strategies* for increasing efficiency of variable costs, especially of indirect variable costs, will target:

- Knowing the exact cost variables for each PFC (usually known for each cost center/process/factory area);
- Identifying as accurately as possible the variable costs related losses—cost savings through value-added analysis (Posteucă, 2018, pp. 35–37; Posteucă and Sakamoto, 2017, pp. 116–127);
- Identification of critical losses related to variable costs (losses causing other losses and waste along all the main processes of each PFC) to develop cost-saving strategy (CSS) in a timely manner;

Depending on the *basic MCI strategies* to increase the efficiency of variable costs, the most credible and feasible assumptions for each basic MCI strategy are determined and *kaizen* and *kaikaku* projects are designed to significantly reduce variable costs to cope with CPV and CPM in one of the four possible business scenarios presented in Figure 2.1.

2.2.2.2 Minimize Inputs by Maximizing Fixed Cost Efficiency

This *MCI change driver* for increasing the efficiency of fixed costs aims to ensure internal manufacturing target profits through minimizing inputs (I) and has the following main purposes:

- Exact location of fixed costs for each main process of each PFC;
- Exact localization of fixed costs losses and waste at the level of each main process of each PFC;
- CLW analysis of each PFC from the fixed costs perspective;
- Continuous reduction of fixed costs for each PFC;
- Increase the productivity of operators in particular;
- Cost reduction of spare parts;

The main system elements related to the increase in efficiency of the possible fixed cost are:

- Direct and indirect labor costs (from the transformation cost structure);
- Cost of direct and indirect labor losses (management losses, motion losses, line organization losses, losses resulting from failure to automate, measuring and adjustment losses)
- Maintenance cost/spare parts cost/ maintenance materials cost (manufacturing overhead; from the transformation cost structure);
- Costs of maintenance losses;
- Depreciation costs (especially equipment; manufacturing overheads; from the transformation cost structure);
- Costs of depreciation losses (the lost opportunity to have an NCU load).

In this context, *basic MCI strategies* for increasing efficiency of fixed costs will target:

- Knowing as exactly as possible the cost of fixed costs for each PFC (usually direct and indirect labor costs are known quite accurately for each cost center/process/factory area);
- Identifying as accurately as possible the fixed cost losses (increase of OEE and elimination of NVAA from process lead time) to develop a cost-saving strategy (CSS) in a timely manner (Posteucă, 2018, pp. 35–37; Posteucă and Sakamoto, 2017, pp. 116–127);

Depending on the *basic MCI strategies* to increase the efficiency of fixed costs, the most credible and feasible assumptions for each MCI strategy are determined, and *kaizen*, MDC and *kaikaku* projects are planned to significantly reduce fixed costs to cope with CPV and CPM in one of the four possible business scenarios presented in Figure 2.1.

2.2.2.3 Minimize Inputs by Continuously Improving Manufacturing Lead Time and Continuously Aligning Manufacturing Processes to Market Needs

This *MCI change driver* for increasing the efficiency of the manufacturing flow aims in particular to ensure internal manufacturing target profits through minimizing inputs (I) and has the following main purposes:

- Decreasing manufacturing lead time by successfully implementing *kaizen*, *kaikaku* and MDC projects;
- Shortening decision times by using information flow procedures;
- Aligning all manufacturing process to takt time (and subsequently of all processes in the company and outside the company);
- Reduce WIP: WIP from Set-up (WIP S) and WIP from Transfer (WIP T);
- Reduce WIP costs;
- Model the manufacturing process as One-Piece Flow (OPF) and Continuous Flow (CF);
- Increase the manufacturing process flexibility by reducing batch size;
- Reduce/eliminate overtime;
- Reduce material conveyances by reducing material handling and distance between equipment or processes.

The main system elements related to manufacturing flow are: (1) number of workstations; (2) bottlenecks; (3) distances between materials and people; (4) transfer times between processes; (5) undefined WIP zones and with undefined quantities (next to equipment or assembly lines); (6) takt time; (7) cycle time; (8) line balancing; man-hours; (9) set-up time; (10) WIP from set-up time; (11) WIP from transfer time; (12) MTTR for buffer inventory reduction; (13) inventory level in the warehouse: parts, components and raw materials; (14) inventory level near equipment and/or line; (15) scrap and rework area; (16) hazardous areas for operators; (17) areas with environmental risk; and (18) OEE, etc.

In this context, *basic MCI strategies* for continuously improving manufacturing lead time and continuously aligning manufacturing processes to market needs (to takt time) shall target:

- Aligning manufacturing lead time to supply lead time and delivery lead time–speed and agility strategy (SAS) for the entire manufacturing flow;
- Prediction and forecast for customer requests (SAS);
- Prediction and forecast for supply requests—continuous alignment of suppliers with constantly reworked manufacturing takt time—decreasing (SAS);
- Manufacturing planning (prevention/reduction/elimination of changes in manufacturing planning; generating systemic waste and losses (SAS));
- Reducing the size of batches by successfully implementing the *kaizen, kaikaku* and MDC (SAS) projects;
- Identification of high WIP zones of WIP from Set-up (WIP S) and WIP from Transfer (WIP T) (SAS);
- Buffer inventory reduction by reducing MTTR (SAS);
- Reduction in number of operations (SAS);
- Reduction/removal of bottlenecks (SAS);
- Reduction of transfer times and related WIP—especially manual transfer (SAS);
- The WIP knowledge and reduction strategy in line with mixed manufacturing (SAS);
- The strategy for establishing the min–max level for WIP in accordance with the takt time, with the actual capacity of the equipment—OEE, depending on the capacity of the internal logistics—the complete lead time of a supply route for the work areas, depending on the standard level of WIP from Set-up (WIP S) and WIP from Transfer (WIP T), according to mixed manufacturing, depending on the changes in the manufacturing plan and according to the standard level of scrap and rework (SAS; impact of effectiveness on efficiency);
- The strategy for the development and implementation of *poka-yoke* (SAS) prevention and/or detection devices (SAS; impact of effectiveness on efficiency);
- Set-up improvement strategy (event number and average time of an event) (SAS);

- Layout redesign strategies to unify assembly lines (if applicable), reduce WIP from Set-up (WIP S) and WIP from Transfer (WIP T) by reducing NVAA, SAS, integrating the new products/processes (SAS; impact of effectiveness on efficiency);
- Classifying all operations in takt time to reduce man-hours—with a takt time decreasing periodically (SAS);
- Balancing all operations of the assembly lines according to the takt time (for industries that have assembly lines) (SAS);
- Managing the bottlenecks and takt time for quality control (SAS);
- Ensuring the flow of office information to support flow material to provide materials on time in full (SAS);
- Ergonomics improvement strategy to support takt time (SAS);
- Reduction of worthless activities and the process lead time (product number increase strategy–PNIS; impact of effectiveness on efficiency);
- Reducing manufacturing lead time by increasing OEE (PNIS; impact of efficiency on efficiency);
- The strategy of designing and implementing new products and new equipment through new and easy-to-process processes with an acceptable level of quality and unit costs (SAS; PNIS; CSS; impact of effectiveness on efficiency);
- Reduction/elimination of shortage of raw materials and materials for equipment and equipment breakdown (PNIS; impact of effectiveness on efficiency);
- Maintenance strategy for tools, dies, jigs (PNIS; impact of effectiveness on efficiency)
- Continuous increase in overall line efficiency (OLE) for PFCs that have the same takt time, regardless of the OLE computation method (PNIS; impact of effectiveness on efficiency);
- Reducing/eliminating the time associated with accidents and occupational diseases (PNIS; impact of effectiveness on efficiency);
- Establishing and improving the new processes of new products so that they fall within a low takt time (PNIS; impact of effectiveness on efficiency);
- Training strategies for all people in and outside the company to: (1) support maintenance system (autonomous maintenance (AM); condition-based maintenance (CBM); time-based maintenance (TBM), etc.), bottleneck reduction, importance of cycle time and factory lead time, quality, WIP, product mix and flexibility,

importance of NVAA, overtime, material handling, safety, environment, MCI/CLW/CCLW, improvement culture, etc.) (SAS; PNIS; CSS);

- Avoiding, reducing and eliminating overtime (SAS; PNIS; CSS);
- Material handling improvement to reduce lead time in manufacturing process (SAS; PNIS; CSS);
- Reducing cycle time for equipment and operators to reduce the need for overtime (SAS; PNIS; CSS);
- Reducing/eliminating the time associated with the lack of knowledge/existence/correct application of the labor procedures by the operators (PNIS; impact of effectiveness on efficiency); and last but not least,
- Continuously identifying all CLWs for each important process of each PFC (CSS);

Depending on the *basic MCI strategies* for continually improving manufacturing lead time and continuously aligning manufacturing processes to market needs (takt time), the most credible and feasible assumptions for each MCI strategy are determined and *kaizen*, *kaikaku* and MDC projects are planned to cope with the CPV and CPM by achieving OMIs and KPIs targets, and in particular the KPIs related to losses and waste at each important process of each PFC, based on the continually redesigned PMP.

2.2.2.4 Minimize Inputs by Continually Improving Inventory Levels: Raw Materials, Components, Consumables, and Finished Products

This *MCI change driver* is related to the increase of efficiency of the inventory (apart from manufacturing flow), but has a major impact on manufacturing efficiency and aims in particular to ensure internal manufacturing targets profit through minimizing inputs (I). It has the following main purposes:

- Ensuring an inventory level that meets the needs of external customers and internal customers (of manufacturing);
- Real-time accurate knowledge and continuous reduction of inventory level (searching to achieve the "zero-stock" principle over the current maximum allowable—standard maximum inventory level for that period based on the manufacturing mix);

- Knowing the cost of inventory level by category (especially for obsolete inventories);
- Keeping the inventory in optimal condition (available for use at any time);
- Continuous sizing and resizing of inventory levels of raw materials, components, consumables, and finished products in order to obtain the best efficiency of the manufacturing process;
- Improvement of raw materials, components, consumables, and finished products lead time and flow;
- Continuous reduction of supply lead time;
- Continuous reduction of delivery lead time;
- Supply lead time alignment at manufacturing takt time;
- Manufacturing lead time alignment at delivery takt time;
- Delivery lead time alignment at customer takt time;
- Knowledge and continuous improvement of inventory levels depending on the manufacturing mix of products;
- Maximizing On Time In Full (OTIF) for the delivery of finished products (ensuring deliveries in accordance with the agreed terms, and that delivery was made in full and the products ordered are not missing);
- Accurate knowledge and continual reduction of inventory level.

The main system elements, those with high potential for change over time and/or disruption of availability, related to the increase in efficiency of the inventory are:

- The possibility of establishing customer requests with acceptable accuracy for 12 months, 90 days, 60 days, 30 days;
- The accuracy of weekly customer orders;
- The possibility of establishing requests to suppliers with acceptable accuracy for 12 months, 90 days, 60 days, 30 days;
- The accuracy of the inventory management system and the real observance of the First In, First Out (FIFO) method;
- The criteria for establishing obsolete inventories for each type of landmark;
- The min–max for each inventory category for each time period according to the mix of the products from the manufacturing;
- The lead time elements at the level of processes of each inventory category (material and information flow elements);
- Truck lead time by category of routes and transport categories.

In this context, *basic MCI strategies* for increasing efficiency of the inventory will target:

- Forecasting customer orders and orders to suppliers;
- maximizing OTIF;
- Reduction of stocks with raw materials (inventory days);
- Reduction of stocks with components (inventory days);
- Reduction of stocks with consumables (auxiliary) (inventory days);
- Reduction of stocks with finished products (inventory days);
- Reducing truck delay (number of events, average time of an event);
- Reducing number of stock-keeping unit (SKU) changes;
- Reducing supply lead time;
- Reduction of delivery lead time;
- Increasing the degree of loading of current warehouses in a rational manner and avoiding the need to invest in additional storage facilities.

Depending on the *basic MCI strategies* to increase efficiency of inventory, the most credible and feasible assumptions for each basic MCI strategy are determined and *kaizen*, *kaikaku* and MDC projects are planned to handle CPV and CPM by achieving OMIs and KPIs targets.

2.3 ESTABLISH THE FRAMEWORK FOR DEVELOPING BASIC MCI STRATEGIES BASED ON THE CURRENT LEVEL OF CLW

In order to achieve a continuous reduction of product unit cost of manufacturing and to ensure a competitive price level, MCI core strategy aims to:

- *Accurately identify the current status of the CLW from the main processes of each PFC,* usually in the proportion of 30%–40% of the current manufacturing costs. The CLW structure is shown in Figure 2.4. Even though the losses and waste levels tend to decrease over time due to systematic and systemic improvements, the current loss and waste level capturing takes place on a continuous basis to capture the current state of non-productivity, and subsequently

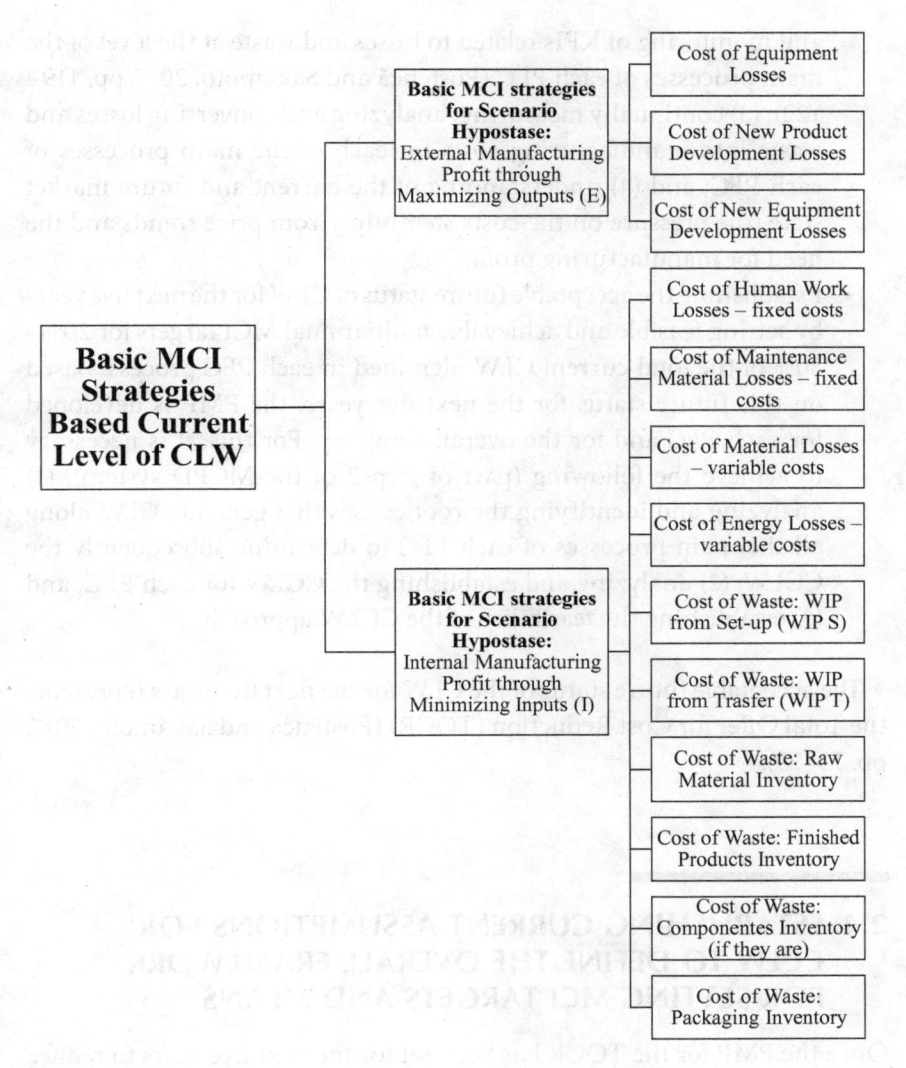

FIGURE 2.4
Framework for developing basic MCI strategies based on the current level of CLW.

the current state of CLW is determined on a scientific and ongoing basis by observing, measuring and experimenting all the events in each PFC process. Limiting the negative influence of uncertainty on achieving the annual and multiannual manufacturing target profits by addressing the current status of CLW (step 1 and part of step 2 of the MCPD system) requires: (1) deep knowledge and understanding of the past and current state of OMIs and KPIs at the level of the main processes of each PFC; (2) continuous development, measurement and monitoring of KPIs related to losses and waste at the level of the main processes of each PFC (Posteucă and Sakamoto, 2017, pp. 119–122); (3) continually measuring, analyzing and converting losses and waste into manufacturing costs for each of the main processes of each PFC; and (4) understanding of the current and future market state (the pressure on the costs stemming from price trends and the need for manufacturing profit);

- Establishing the acceptable future status of CLW for the next five years by setting feasible and achievable multiannual MCI targets for 20%–50% of the total current CLW identified in each PFC process. Based on this future status for the next five years, the PMP is developed for each PFC and for the overall company. For this, it is necessary to achieve the following (part of step 2 of the MCPD system): (1) analyzing and identifying the root causes that generate CLW along all the main processes of each PFC to determine subsequently the CCLW; (2) analyzing and establishing the CCLW for each PFC; and (3) establishing the feasibility of the CCLW approach.

The acceptable future status of the CLW for the next five years represents the Total Offer for Cost Reduction (TOCR) (Posteucă and Sakamoto, 2017, pp. 73–75).

2.4 ESTABLISHING CURRENT ASSUMPTIONS FOR CCLW TO DEFINE THE OVERALL FRAMEWORK FOR SETTING MCI TARGETS AND MEANS

Once the PMP for the TOCR has been set for the next five years to reduce the CLW, the annual CLW level that can be addressed by reducing the annual CCLW level at each PFC level is determined.

By developing current assumptions for CCLW, especially credible assumptions, it is sought to identify and consistently address the main phenomena, principles and symptoms of the CLW, even though their structure may have significant fluctuations from one period to the next, in particular, depending on the product types manufactured in the processes of each PFC. With all these fluctuations, for example, a *kaizen* project to reduce breakdown equipment losses, the cost of waste and waste attracted throughout the entire manufacturing flow can be considered necessary almost without taking account of the types of products that will be made in the future. This *kaizen* project for CCLW reduction can be deemed to be one of the credible assumptions for fulfilling MCI strategy for that year. By identifying the CCLW that is easy to address (easy to implement, feasible, and consistent over time), it is sought to mitigate the impact of uncertainties related to the evolution of sales on the manufacturing profit and the scientific establishment of MCI targets and means. Reducing the CCLW annual level for each PFC is achieved by setting MCI targets and means for each CLW that has a behavior that causes other losses and waste along the entire manufacturing flow—in upstream and downstream processes relative to the process where the CCLW underlying phenomenon occurs; for example, a process in which the equipment generates rework.

As shown in Figure 2.1, in the early years of application of the MCPD system, starting from the multiannual and annual situation of the need for manufacturing profit ("E" and "I") of a manufacturing company for a PFC or for the whole company, setting the annual MCI target and means—through continuous reconciliation between the required level of product unit cost of manufacturing imposed by the need for price competitiveness, the need for annual and multiannual profit, and the real possibilities to achieve the MCI targets—for one of the four annual scenarios can be:

- *Scenario 1—Major and flexible sales growth:* to achieve notable progress in volume and structure of products, usually planned, and to reach the annual and multiannual MCI targets according to internal and, especially, external needs by implementing MCI means (*kaizen* and *kaikaku*); the annual percentage of MCI targets varies between 6% and 10% of the average product unit cost of manufacturing and is mainly achieved through *kaizen* and *kaikaku* projects. For example: if the total predicted spending of a PFC is $10,000,000, the identified CLWs vary between 30%–40% at PFC processes ($3,000,000–$4,000,000 respectively), the next acceptable state of CLW for the next five years (or TOCR) varies between 20%–50% of the identified CLW (i.e. $600,000–$2,000,000 CCLW), then the annual MCI target for Scenario 1 varies between 6%–10% of the average product unit cost of manufacturing (or $600,000–$1,000,000 of the TORC). This annual stake of reducing average product unit cost of manufacturing by reducing TORC is the profit plan achieved by increasing the annual productivity of all the main PFC processes, in particular by accentuating the change drivers of MCI and the elements related to change drivers of MCI for the scenarios hypostases of external manufacturing profit through maximizing outputs ("E"). For this scenario, the 6%–13% annual level of MCI targets is predominantly established depending on the market signals (the actions of competitors, customers and suppliers on current and/or new product volumes and their pricing policies), but also according to their own actions (new products launched, new technologies implemented, etc.) and the level of resources allocated to improvements (MCI means).

- *Scenario 2—Incremental growth*: to achieve predetermined incremental product volumes and product structure and to achieve annual MCI targets of at least 6% of the average product unit cost of manufacturing by implementing MCI means (especially *kaizen* and *kaikaku*). The focus on annual change drivers of MCI and the elements related to change drivers of MCI is balanced, both for scenario hypostases of internal manufacturing profit through minimizing inputs ("I") and external manufacturing profit through maximizing outputs ("E"). In the case of Scenario 2, which is the most common scenario, after the first three to five years of consistent application of the MCPD system, the annual MCI target levels may drop to 3%–5% of the average product cost of manufacturing amid the reduction in TORC opportunities.
- *Scenario 3—Incremental decline*: to cope with relatively small and constant decreases in product volume and structure and to avoid a disaster in terms of annual and, especially, multiannual manufacturing profit; with a significant impact on the current structure of CLW; with the need to make annual improvements to MCI (especially *kaizen* and *kaikaku*) by setting annual MCI targets of 6%–8% of the average product unit cost of manufacturing, in particular by emphasizing the change drivers of MCI and elements related to change drivers of MCI-related scenario hypostases of internal manufacturing profit through minimizing inputs ("I") to prepare for launching new products and new technologies, in particular, by preparing the emphasizing of the scenario hypostases of external manufacturing profit through maximizing outputs ("E").
- *Scenario 4—Major unexpected changes*: to cope with or react quickly to major external and internal changes in the volume and structure of products; with a major impact on the current structure of the

CLW; with the need to make improvements to MCI, especially by *kaikaku*. In the case of this scenario, the annual MCI target level is 6%–10% of the average product unit cost of manufacturing, in particular through major concerns about the change drivers of MCI and the elements related to change drivers of MCI related to scenarios hypostasis of internal manufacturing profit through minimizing inputs ("I"). The main objective is to ensure the continuity of the manufacturing company's activities.

Therefore, all the work of developing structured scenarios to ensure the continuous and healthy transformation of manufacturing flow for each PFC through the MCPD system, presented in Figure 2.1, aims to establish the framework for setting annual MCI targets and means to support scientifically the desire of all people in and outside the company to participate continually and pro-actively in achieving productivity improvements. The detailed setting of the annual MCI targets and means at the level of each PFC process will be presented in the next chapter.

2.5 THE TANGIBLE AND INTANGIBLE EFFECTS OF CONTINUOUS AND CONSISTENT APPLICATION OF THE MCPD SYSTEM

Continuous support for the perennial results of manufacturing companies at the level of turnover, operating profit, product number converging to market share, quality, delivery and safety and morale to support CPV and CPM requires the development of PMP for each PFC and for the overall company, irrespective of whether scenarios hypostases are stressing the role of external manufacturing profit through maximizing outputs ("E") or increasing the role of internal manufacturing profit through minimizing inputs ("I") and whether sales are rising or declining in one period or another.

The PMP design will aim to continuously support the eight MCI change drivers and MCI system elements by developing MCI strategies to achieve annual and multiannual MCI targets and means through continuous improvement of CCLW and implicitly of CLW.

The main tangible effects of the MCPD system, tracked and reached at the level of the OMIs, KPIs and DMIs, are the change drivers of MCI:

- *CD1—Effectiveness of current equipment ("E")*: product number increase by increasing OEE and OLE; determining the optimal moment for a new equipment (purchase/development);
- *CD2—Effectiveness of new equipment ("E")*: reducing the cycle time of new equipment; reducing the set-up time of new equipment and associated WIP set-up; decrease man-hours per product; scrap reduction; reducing the cost of investing in new equipment by scientifically determining the type of equipment needed and the Life Cycle Cost (LCC); reducing Time-to-Start Manufacturing (TtSP)
- *CD3—Development of new profitable products ("E")*: reducing material costs; reducing CLW after launching new products; increasing the number of hybrid components and raw materials (used in many products); reducing utility consumption; reducing scrap and rework; increasing capacity utilization; reducing Time-to-Market (TtM), etc. (Posteucă and Zapciu, 2015a);
- *CD4—Maximizing variable cost efficiency ("I")*: reducing raw material costs; reducing auxiliary materials costs; reducing energy costs; reducing manufacturing overheads–variable part; reducing tool costs; reducing die and jig costs;
- *CD5—Maximizing fixed cost efficiency ("I")*: direct and indirect labor costs reduction; reducing maintenance costs/spare parts costs; reducing depreciation costs;
- *CD6—Continuously improving manufacturing lead time ("I")*: reducing manufacturing lead time (information and material); reducing Set-up time; reducing WIP from Set-up (WIP S); reducing WIP from Transfer (WIP T); reducing internal logistics time (material handling); increasing the speed of assembly lines (if applicable); reducing defective parts per million (DPPM), reducing rework; reducing health and safety incidents, etc. (Posteucă and Zapciu, 2013; Posteucă and Zapciu, 2015d);
- *CD7—Continuously aligning processes to market needs ("I")*: reducing cycle time; reducing total value-adding time (VAT) or Non-Value-Added Activities (NVAA); reducing bottlenecks;
- *CD8—Continually improving inventory levels ("I")*: increase on time delivery (OTD); production delivery performance (OTIF); reducing truck delays (as average); reducing raw material inventory days; reducing finished product inventory days; reducing components inventory days; reducing packaging inventory days (Posteucă and Zapciu, 2015c).

All these tangible effects are found at the level of the continuous reduction of the average product cost of manufacturing imposed by the market (market share and manufacturing target profit) and, implicitly, at the level of the CLW reduction for:

1. Cost of Equipment Losses;
2. Cost of New Equipment Losses;
3. Cost of New Product Development Losses;
4. Cost of Human Work Losses;
5. Cost of Material/Auxiliary Material Losses;
6. Cost of Maintenance Material Losses;
7. Cost of Energy Losses;
8. Cost of Internal Logistic Losses (assimilated to Equipment Losses);
9. Cost of Waste—WIP from Set-up (WIP S);
10. Cost of Waste—WIP from Transfer (WIP T);
11. Cost of Waste—Raw Material Inventory;
12. Cost of Waste—Finished Products Inventory;
13. Cost of Waste—Components Inventory and
14. Cost of Waste—Packaging Inventory (Posteucă and Zapciu, 2015b).

Over time, the following intangible effects obtained by the continuous and consistent application of the MCPD system are found: (1) increasing people's trust in the company in the MCPD system; (2) increasing satisfaction for scientific planning of productivity improvements by involving all people; (3) improving teamwork; (4) continuous improvement of people's knowledge about the manufacturing flow of each PFC; (5) continuous improvement of contextual managerial behavior (Posteucă, 2011).

2.6 CONCLUSIONS

In order to transpose the manufacturing unit cost reduction strategy into MCI's annual activities and actions, it is necessary to take into account the sales trends in order to require productivity improvements to provide the appropriate methods, techniques and tools in the two possible cases: (1) maximizing efficiency for a particular PFC if the trend of current and future product sales is potentially rising; and otherwise (2) primarily maximizing efficiency for a particular PFC if the trend in current and future product sales is potentially falling. The establishment and implementation of annual MCI targets and means for each PFC can be achieved predominantly by concentrating on and achieving the annual external production target profit or by concentrating on and achieving the annual internal manufacturing target profit.

The approach of MCI change drivers and associated system elements aims to develop a MCI-based strategy to ensure harmonious transformation, continuously linked to market and shareholder needs, of the whole manufacturing system by fundamentally continuing changing for the better the method of work or the way to see things by successfully selecting and implementing profitable systematic and systemic improvement projects by approaching the current level of CLW and CCLW to materialize the annual MCI goal by reaching annual MCI targets.

Therefore, by the continuous reduction of the CLW, month by month and year by year, both the increase of the external manufacturing profit through the maximizing outputs ("E") and the internal manufacturing profit through minimizing inputs ("I") are enhanced, regardless of the increasing or decreasing trend, current or prospective, of sales by targeting productivity improvements to manufacturing target profits. To be convergent to CPV and CPM, to reach the annual MCI goal on an annual basis represents the core concern of the MCPD system.

3

The MCPD Transformation: Establishing and Implementing Profitable Improvement Projects

Compared to the previous presentations of the MCPD system structure, from the already published material, this chapter will emphasize the mechanism for choosing the most profitable productivity improvement projects, projects chosen to achieve MCI consistently. The basis of this mechanism is to set and achieve the annual MCI targets, aiming at achieving a continuous and harmonious transformation of the entire manufacturing flow of each PFC and the entire company, depending on the market signals, according to the need for manufacturing profit— constant profit to ensure the existence and development of the company— and according to the current state of the company's main processes. At the same time, it will address the method of implementing profitable productivity improvement projects to achieve the annual MCI targets.

Therefore Section 3.1 will present the main ingredients of selecting profitable improvement projects. Sections 3.2, 3.3 and 3.4 show the three phases and seven steps of the MCPD system. The first two steps of the MCPD system will be addressed in more detail to show the method of establishing MCI targets and means (MCI policy) and, implicitly, the method of choosing the most profitable improvement projects. Section 3.5 outlines the conclusions of this chapter.

The chapters of the second and third parts of this book present examples of the implementation of systematic and systemic improvement projects for the achievement of the annual MCI goal.

3.1 MARKET AND PROFIT DRIVEN FOR ANNUAL MCI GOAL: FROM COMPANY PRODUCTIVITY VISION TO ANNUAL PROFITABLE IMPROVEMENT PROJECTS

From the perspective of the MCPD system, multiannual manufacturing target profit, which is continually converging to CPV and CPM, is established and achieved by both the multiannual external manufacturing profit and by the multiannual internal manufacturing profit contribution. At the annual level, the annual manufacturing target profit is set and achieved by both the annual manufacturing external profit (from sales) and by the annual MCI goal, or by achieving the objective of reducing CLW, namely CCLW, by establishing and achieving the annual MCI targets and means at the level of the main processes of each PFC. In this way, the annual MCI goal will be set and achieved at the level of each PFC, at the level of each process of each PFC, especially at the company level, for all PFCs. Therefore, the year after year achievement of the annual MCI goal, as a core concern of the MCPD system, requires the continued assumption of market pressures and requirements at the level of the main processes of each PFC. These contextual and market-specific pressures and constraints for each PFC require alignment in terms of the likely level of annual MCI goal achieved in terms of:

- The number of products it is possible to manufacture and sell for each market segment;
- The likely average price level of the products;
- The mix of current and future products within each PFC;
- The life cycle of current and future products;
- Orders from customers unfulfilled in the past;
- New products needed to be designed and launched to align the product mix to the competitive level of other companies on the market; and
- Supplier trends and behaviors.

For the annual alignment of the MCI goal on the market—or market-driven for annual MCI goal—for each PFC there are successive simulations to reconcile the seven main inputs above to determine the most credible assumptions for each annual MCI strategy and for the subsequent establishment of annual MCI targets and means (see Table 2.1).

Starting from the market-driven for annual MCI goal, the profit driven for annual MCI goal is determined. Achieving the annual manufacturing target by achieving the annual MCI goal, for the overall company, for each PFC, and possibly for each product, requires identification of the process-based sources of profit participation of CLW and CCLW. Once again, through successive simulations to identify these sources of participation in achieving the annual manufacturing target profit by achieving the annual MCI goal, the antagonistic pressures of expectation for manufacturing profit in the medium-term and the long-term are reconciled with the real possibility of sustaining the annual MCI goal by addressing the opportunities and challenges of CCLW at the level of the main processes of each PFC. Establishing annual MCI targets (driven by the annual MCI targets development) requires a deep knowledge and mastery of the manufacturing flow from the perspective of the MCPD system, i.e. a deep knowledge of losses and waste, costing, budgeting, and the CLW determination system and the CCLW for each process of each PFC.

Therefore, choosing and implementing the most profitable productivity improvement projects that are continually converging to the CPV depends on the accuracy of the data and information relating to the costs provided by the cost and managerial accounting, and the continuous capture of losses and waste from processes, or the capture of non-productivity and transposing this non-productivity into costs by determining the CLW with the highest accuracy (Posteucă, 2018, p. 81–138).

Continuous exploitation of the CLW by setting and achieving the MCI targets to achieve MCI on a continuous basis represents in fact *uncovering and exploiting the hidden reserves of manufacturing profit.*

3.2 PHASE 1: MANUFACTURING COST POLICY ANALYSIS

The goal of manufacturing cost policy analysis, the first phase of MCPD, is *to broadly define the issue of need for MCI* based on a deep understanding of past, current and future business conditions, and to identify the way in which the CLW manifests by identifying the root cause of the CCLW (*Step 1: context and purpose of MCI*) and *to set countermeasures against the root cause of CCLW* by choosing the most profitable systematic (*kaizen*) and systemic (*kaikaku*) improvement projects of productivity and the best moments to address an MCI problem at the shop floor level through problem-solving techniques (PST). In fact, it aims to align MCI annually with the need to achieve the annual MCI targets and, furthermore, the need to achieve the annual MCI targets (*Step 2: annual MCI targets and means*).

This first phase of the MCPD system translates the need to reduce the product unit cost of manufacturing into concrete actions at the level of the main processes of each PFC and to increase the real managerial commitment and reduce the resistance to change among all people both in and outside the company.

3.2.1 Step 1: Context and Purpose of MCI

The first step of the MCPD system is to develop and implement both productivity policy deployment to achieve CPV translation to annual activities, and MCI policy deployments to set an annual MCI goal or annual productivity stake. MCI policy deployment aims to increase the number of products sold and decrease manufacturing costs. Productivity policy deployment and MCI policy deployment will be detailed henceforth.

3.2.1.1 Productivity Policy Deployment: from Company Productivity Vision to Annual Activities and Actions

In order to achieve productivity policy implementation as a systematic planning process for a manufacturing company, to align MCI scenarios and strategy with day-to-day business operations and ongoing productivity improvements, it is necessary to detail and communicate CPV and CPM across all people in the company so that they can continually be aware of their role and purpose in achieving the CPV and CPM.

Figure 3.1 summarizes a *productivity policy deployment*, which is in fact a manufacturing company policy deployment. It addresses:

- *Productivity policy* or direction of action proposed by an organization (through targets and means); and
- *Productivity deployment* or positioning of all resources ("combat troops") in action ("military action") to carry out effective and efficient actions and activities to achieve targets through means (Akao, 1991; Wood and Munshi, 1991).

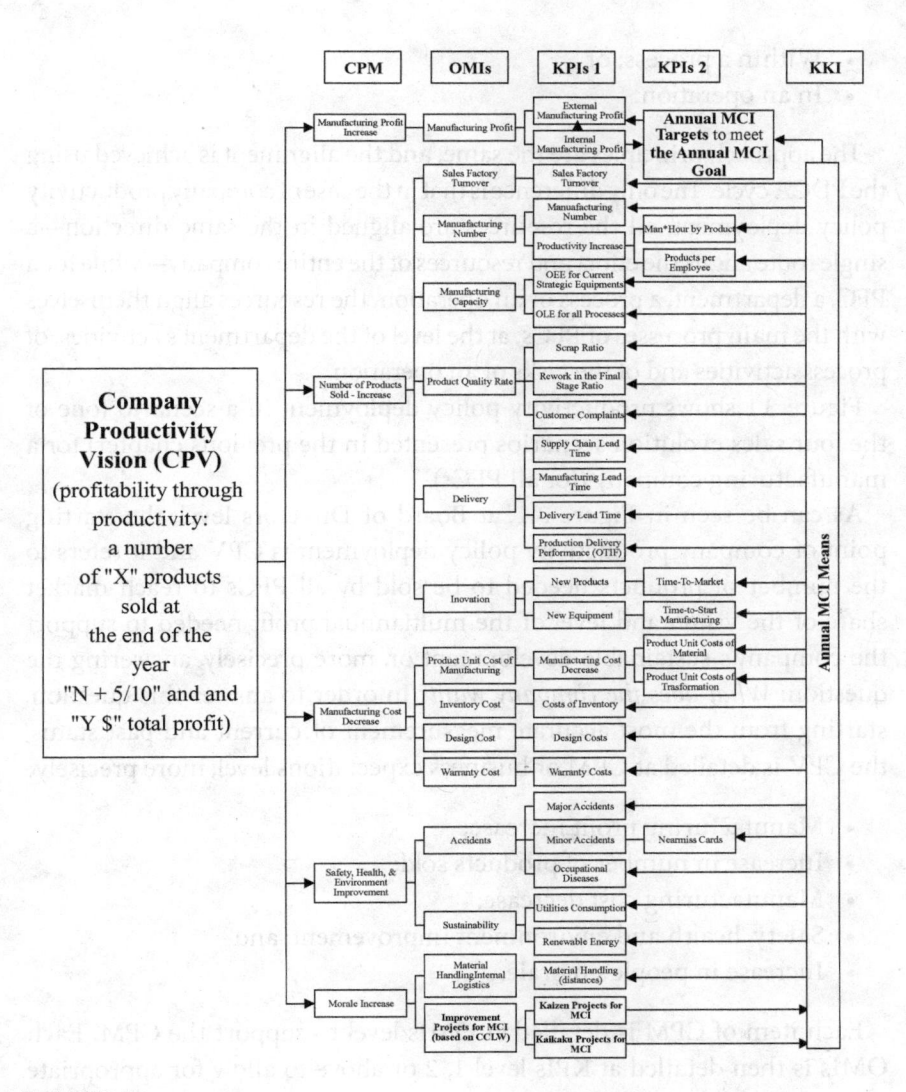

FIGURE 3.1
Productivity policy deployment: from company productivity vision to annual activities and actions.

Productivity policy deployment can be applied:

- At the level of an entire manufacturing company, for all PFCs, at the top of an organization and downwardly deployed through each function, section or process;
- At the level of a PFC;
- Within a department;

- Within a process; or
- In an operation.

The approach principles are the same, and the alignment is achieved using the PDCA cycle. The only difference is that in the case of company productivity policy deployment, all the resources are aligned in the same direction—a single route aligns the effort and resources of the entire company—while for a PFC, a department, a process or an operation, the resources align themselves with the main processes of PFCs, at the level of the department's activities, of process activities and of the tasks of an operation.

Figure 3.1 shows productivity policy deployment in a scenario (one of the four sales evolution scenarios presented in the previous chapter) for a manufacturing company (for all PFCs).

As can be seen in Figure 3.1, at Board of Directors level, the starting point of company productivity policy deployment is CPV and it refers to the number of products needed to be sold by all PFCs to reach market share of the vision and level of the multiannual profit needed to support the company's sustainable development or, more precisely, answering the question: *What does the company want?* In order to answer this question, starting from the most accurate measurement of current and past status, the CPV is detailed at CPM or business expectations level, more precisely:

- Manufacturing profit increase;
- Increase in number of products sold;
- Manufacturing cost decrease;
- Safety, health and environment improvement; and
- Increase in people's morale.

Each item of CPM is detailed at OMIs level to support the CPM. Each OMIs is then detailed at KPIs level 1, 2 or above to allow for appropriate process or system level measurements and to set out improvement targets based on current productivity improvement opportunities. From the perspective of the MCPD system, the setting of productivity improvement targets is based on annual CCLW targets (annual credible assumptions for each MCI strategy; see Figure 2.1), with *kaizen* and *kaikaku* projects for MCI or, more precisely, with the help of annual strategic MCI (by meeting KKI targets) to achieve annual MCI targets and, implicitly, the annual MCI goal (Posteucă and Sakamoto, 2017, pp. 9–66 and pp. 142–154). Setting the annual MCI goal is based on the need to achieve the product unit cost of manufacturing decrease (Posteucă, 2018, p. 15).

Therefore, the answer to the initial question of the Board of Directors (*What does the company want?*) aims, in particular, for customer satisfaction (cost, quality and delivery times) and shareholder satisfaction or shareholder value (in particular, market share and profit).

As can be seen in Figure 3.1, annual and multiannual manufacturing profitability through external and internal manufacturing profit focuses on the two main directions of the CPM: (1) number of products sold (which must support the increase in sales volumes and, implicitly, the increase in sales revenue) and (2) manufacturing cost decrease (which must support the continuous decrease of unit sales prices of products and, implicitly, the level of spending in the manufacturing area). Safety, health, and environmental improvement and people's morale increase are strategic elements that support the continuity of the company's business. Costs for their improvement, as well as other improvements, are considered in the AMIB and AMCIB sizing and, implicitly, in the setting of annual MCI targets.

Furthermore, from the perspective of the MCPD system, to continuously support the answer to the question (*What does the company want?*), the CLW is determined and the CCLW is established to meet the annual MCI targets and, implicitly, the annual MCI goal by achieving annual MCI mean targets. The establishment of annual MCI targets and means (MCI policy) requires continuous, detailed and accurate measurement of generic KPIs and KPIs related to losses and waste at the level of each PFC process.

Therefore, an annual reconciliation between MCI targets and MCI means is required based on the deep understanding of the manufacturing flow, starting from the focus on current and future market needs and on the current and future manufacturing capacity and the need to define the ideal and future state of manufacturing flow (Posteucă, 2018, pp. 95–130).

Table 3.1 presents a first example of company productivity policy deployment (CPM: number of products sold; OMI: manufacturing capacity; KPIs1: OEE for current strategic equipment; a section from Figure 3.1).

Table 3.2 presents a second example of company productivity policy deployment (CPM: number of products sold; OMI: delivery; a section from Figure 3.1).

Similarly, this is done for all the CPMs in Figure 3.1.

Therefore, in order to achieve company productivity policy deployment, a deep understanding of the manufacturing flow is required from the perspective of the MCPD system (Posteucă, 2018, pp. 96–116) and setting strategic key points on manufacturing process (SKPMP) (Posteucă, 2018,

TABLE 3.1

An Example of Company Productivity Policy Deployment for OEE's Current Strategic Equipment (in Synthesis)

Company Productivity Vision	CPM	OMIs	KPIs1	KPIs2	KPIs3	Units (**)	"N-3" How much? fill in	"N-3" Where? Processes/System (*)	"N-2" How much? fill in	"N-2" Where? Processes/System (*)	"N-1" How much? fill in	"N-1" Where? Processes/System (*)	"N°" (Target) How much? fill in	"N°" (Target) Where? Processes/System (*)	"N+1°" (Target) How much? fill in	"N+1°" (Target) Where? Processes/System (*)	"N+2/10°" (Target) How much? fill in	"N+2/10°" (Target) Where? Processes/System (*)
	Number of Products Sold - Increase					pieces	x	P/S (***)	x	P/S	x	P/S	x	P/S	x	P/S	x	P/S
		Delivery																
			Supply Chain Lead Time			days	x	P/S	x	P/S	x	P/S	x	P/S	x	P/S	x	P/S
				Overall Material Inventory Days		days	x	P/S	x	P/S	x	P/S	x	P/S	x	P/S	x	P/S
					Raw Materials Inventory Days	days	x	P/S	x	P/S	x	P/S	x	P/S	x	P/S	x	P/S
					Components Inventory Days	days	x	P/S	x	P/S	x	P/S	x	P/S	x	P/S	x	P/S
					Manufacturing Lead Time	days	x	P/S	x	P/S	x	P/S	x	P/S	x	P/S	x	P/S

(Continued)

TABLE 3.1 (CONTINUED)

An Example of Company Productivity Policy Deployment for OEE's Current Strategic Equipment (in Synthesis)

Company Productivity Vision	CPM	OMIs	KPIs1	KPIs2	KPIs3	Units (**)	"N-3" How much? fill in	Where? Processes/System (*)	"N-2" How much? fill in	Where? Processes/System (*)	"N-1" How much? fill in	Where? Processes/System (*)	"N" (Target) How much? fill in	Where? Processes/System (*)	"N+1" (Target) How much? fill in	Where? Processes/System (*)	"N+2/10" (Target) How much? fill in	Where? Processes/System (*)
				WIP Inventory Days		days	x	P/S	x	P/S	x	P/S	x	P/S	x	P/S	x	P/S
					WIP Standard near equipment/lines	days	x	P/S	x	P/S	x	P/S	x	P/S	x	P/S	x	P/S
					WIP from Set-up (on areas)	days	x	P/S	x	P/S	x	P/S	x	P/S	x	P/S	x	P/S
					WIP from Transfer (on areas)	days	x	P/S	x	P/S	x	P/S	x	P/S	x	P/S	x	P/S
				Total Cycle Time		Sec.	x	P/S	x	P/S	x	P/S	x	P/S	x	P/S	x	P/S
					Number of workstations	No.	x	P/S	x	P/S	x	P/S	x	P/S	x	P/S	x	P/S
			Delivery Lead Time			days	x	P/S	x	P/S	x	P/S	x	P/S	x	P/S	x	P/S
				Finished Product Inventory Days		days	x	P/S	x	P/S	x	P/S	x	P/S	x	P/S	x	P/S

(Continued)

TABLE 3.1 (CONTINUED)

An Example of Company Productivity Policy Deployment for OEE's Current Strategic Equipment (in Synthesis)

Company Produc-tivity Vision	CPM	OMIs	KPIs1	KPIs2	KPIs3	Units (**)	"N-3" How much? fill in	Where? Processes/System (*)	"N-2" How much? fill in	Where? Processes/System (*)	"N-1" How much? fill in	Where? Processes/System (*)	"N" (Target) How much? fill in	Where? Processes/System (*)	"N+1" (Target) How much? fill in	Where? Processes/System (*)	"N+2/10" (Target) How much? fill in	Where? Processes/System (*)
					Number of delayed trucks	no.	x	P/S	x	P/S	x	P/S	x	P/S	x	P/S	x	P/S
					Stand-by Finished products ratio	%	x	P/S	x	P/S	x	P/S	x	P/S	x	P/S	x	P/S
					Packaging Inventory Days	days	x	P/S	x	P/S	x	P/S	x	P/S	x	P/S	x	P/S
			Production Delivery Performance (OTIF)			%	x	P/S	x	P/S	x	P/S	x	P/S	x	P/S	x	P/S

Note (*): processes: the main processes of each PFC; system: manufacturing system approach; (**): units of measurement are determined for all PFCs by setting an average of 5–15 representative products for each PFC; (***): P/S: Process/System (manufacturing system).

TABLE 3.2

An Example of Company Productivity Policy Deployment for Delivery (in Synthesis)

Company Productivity Vision						Units (**)	"N-3"		"N-2"		"N-1"		"N" (Target)		"N+1" (Target)		"N+2/10" (Target)	
CPM	OMIs	KPIs1	KPIs2	KPIs3			*How much? fill in*	Where? Processes/System (*)	*How much? fill in*	Where? Processes/System (*)	*How much? fill in*	Where? Processes/System (*)	*How much? fill in*	Where? Processes/System (*)	*How much? fill in*	Where? Processes/System (*)	*How much? fill in*	Where? Processes/System (*)
Number of Products Sold - Increase						pieces	x	P/S (***)	x	P/S	x	P/S	x	P/S	x	P/S	x	P/S
	Delivery																	
		Supply Chain Lead Time				days	x	P/S	x	P/S	x	P/S	x	P/S	x	P/S	x	P/S
						days	x	P/S	x	P/S			x	P/S	x	P/S	x	P/S
			Overall Material Inventory Days			days	x	P/S	x	P/S	x	P/S	x	P/S	x	P/S	x	P/S
				Raw Materials Inventory Days		days	x	P/S	x	P/S	x	P/S	x	P/S	x	P/S	x	P/S
				Components Inventory Days		days	x	P/S	x	P/S	x	P/S	x	P/S	x	P/S	x	P/S
		Manufacturing Lead Time					x	P/S	x	P/S	x	P/S	x	P/S	x	P/S	x	P/S

Years

(Continued)

TABLE 3.2 (CONTINUED)

An Example of Company Productivity Policy Deployment for Delivery (in Synthesis)

Company Productivity Vision	CPM	OMIs	KPIs1	KPIs2	KPIs3	Units (**)	"N-3" How much? fill in	Where? Processes/ System (*)	"N-2" How much? fill in	Where? Processes/ System (*)	"N-1" How much? fill in	Where? Processes/ System (*)	"N" (Target) How much? fill in	Where? Processes/ System (*)	"N+1" (Target) How much? fill in	Where? Processes/ System (*)	"N+2/10" (Target) How much? fill in	Where? Processes/ System (*)
					WIP Inventory Days	days	x	P/S	x	P/S	x	P/S	x	P/S	x	P/S	x	P/S
					WIP Standard near equipment/ lines	days	x	P/S	x	P/S	x	P/S	x	P/S	x	P/S	x	P/S
					WIP from Set-up (on areas)	days	x	P/S	x	P/S	x	P/S	x	P/S	x	P/S	x	P/S
					WIP from Transfer (on areas)	days	x	P/S	x	P/S	x	P/S	x	P/S	x	P/S	x	P/S
				Total Cycle Time		Sec.	x	P/S	x	P/S	x	P/S	x	P/S	x	P/S	x	P/S
					Number of workstations	No.	x	P/S	x	P/S	x	P/S	x	P/S	x	P/S	x	P/S
			Delivery Lead Time			days	x	P/S	x	P/S	x	P/S	x	P/S	x	P/S	x	P/S
				Finished Product Inventory Days		days	x	P/S	x	P/S	x	P/S	x	P/S	x	P/S	x	P/S

(Continued)

TABLE 3.2 (CONTINUED)

An Example of Company Productivity Policy Deployment for Delivery (in Synthesis)

Company Produc-tivity Vision	CPM	OMIs	KPIs1	KPIs2	KPIs3	Units (**)	"N-3" How much? fill in	"N-3" Where? Processes/System (*)	"N-2" How much? fill in	"N-2" Where? Processes/System (*)	"N-1" How much? fill in	"N-1" Where? Processes/System (*)	"N" (Target) How much? fill in	"N" (Target) Where? Processes/System (*)	"N+1" (Target) How much? fill in	"N+1" (Target) Where? Processes/System (*)	"N+2/10" (Target) How much? fill in	"N+2/10" (Target) Where? Processes/System (*)
					Number of delayed trucks	no.	x	P/S	x	P/S	x	P/S	x	P/S	x	P/S	x	P/S
					Stand-by Finished products ratio	%	x	P/S	x	P/S	x	P/S	x	P/S	x	P/S	x	P/S
					Packaging Inventory Days	days	x	P/S	x	P/S	x	P/S	x	P/S	x	P/S	x	P/S
			Production Delivery Perfor-mance (OTIF)			%	x	P/S	x	P/S	x	P/S	x	P/S	x	P/S	x	P/S

Note (*): processes: the main processes of each PFC; system: manufacturing system approach; (**): units of measurement are determined for all PFCs by setting an average of 5–15 representative products for each PFC; (***): P/S: Process/System (manufacturing system).

pp. 141–146; Posteucă and Sakamoto, 2017, pp. 105–107) at the level of each PFC. The continued connection between OMIs and KPIs in the main processes of each PFC is essential. By setting CLW and CCLW at the level of each process of each PFC, annual MCI targets can be established, and then the annual MCI means to achieve the annual MCI goal.

As can be seen in Tables 3.1 and 3.2, the starting point of the targeting process for year "N" is the current state of year "N-1" and the trend of previous years. Then, as will be shown, the targets for year "N" and for subsequent years are set according to:

- The annual MCI goal level and annual MCI targets at the level of the main processes of each PFCs based on annual CCLW through annual MCI means (*kaizen* and *kaikaku* for MCI); and
- Other strategic interests, especially those related to product quality and time levels (cycle time and/or process lead time to meet the takt time).

Principle 1 of the MCPD system has to be observed first and foremost (*annual and multiannual manufacturing target profit from MCI does not change at the level of the whole company*), regardless of the plan or path chosen. Consequently, first the strategic need for annual manufacturing profit is ensured, and then the strategic needs for quality and delivery times. At the annual MCI targets level some of the needs for increasing quality and decreasing process times are already included, but sometimes there are additional requirements from managers based on the actual context of the business at that time and/or the near future.

3.2.1.2 MCI Policy Deployment to Establish Annual MCI Goal: Annual Productivity Stake

As shown in Figure 3.1, in order to establish annual MCI means to support annual MCI targets and, implicitly, the annual MCI goal, it is necessary to know the CLW as accurately as possible in the main processes of each PFC and, implicitly, to determine credible assumptions for each MCI strategy (see Figure 2.1).

Therefore, after answering the question *What does the company want?* (volume of products and MCI) and the question *where are the directions of accomplishing what it is desired* (at the level of the main processes of each

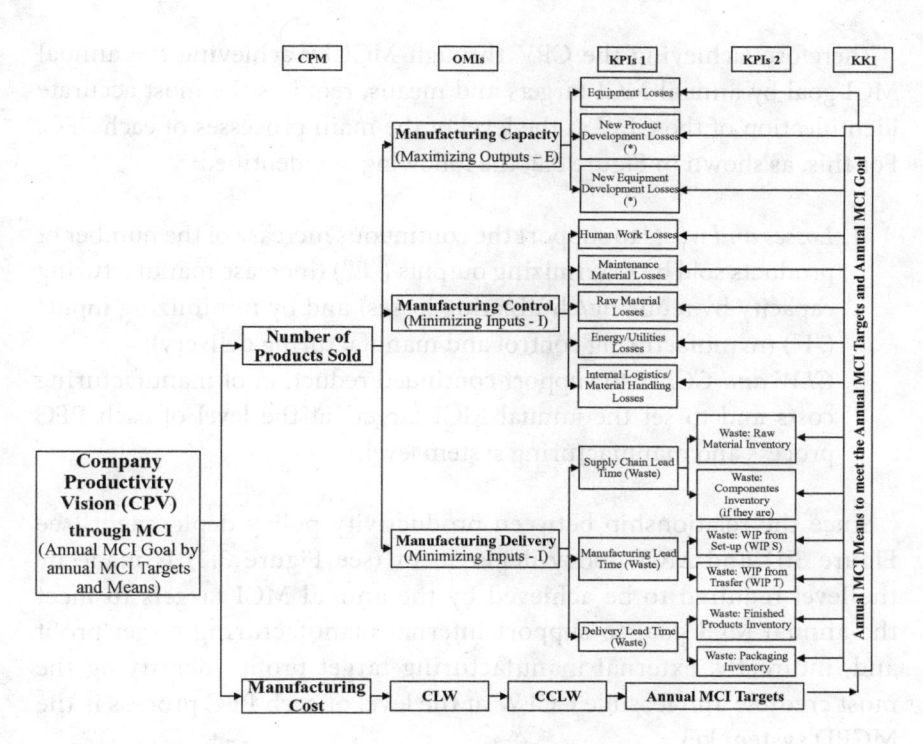

FIGURE 3.2
MCI policy deployment for a PFC for strategically targeting productivity improvement needs.

PFC) in the previous section (Figure 3.1 and Table 3.1), Figure 3.2 presents MCI policy deployment for a PFC to answer the question of *how much* of the annual manufacturing profit can be achieved, namely, what is the basic mechanism for setting annual MCI targets and means.

Note (*): new product development losses and *new equipment development losses* refer to the efficiency and effectiveness losses of both the R&D activities and, in particular, the post-launching and new equipment installation (in processes).

As previously reported in Section 2.4, the annual percentages of MCI targets for the four sales scenarios range from 6% to 10% (although sometimes less than 6% and sometimes more than 10%). These percentages are intended to support the annual and multiannual manufacturing profit plan, a continuous convergent plan with CPV and CPM, through a competitive level of the unitary price, and the increasing volume of products manufactured and sold. These annual percentages, needs originating outside the company, are the ones triggering in the MCPD system the need for annual MCI targets and means.

Therefore, achieving the CPV through MCI by achieving the annual MCI goal by annual MCI targets and means, requires the most accurate identification of the CLW at the level of the main processes of each PFC. For this, as shown in Figure 3.2, the following are identified:

1. *Losses and waste* to support the continuous increase of the number of products sold by maximizing outputs ("E") (increase manufacturing capacity by reducing/eliminating losses) and by minimizing inputs ("I") (manufacturing control and manufacturing delivery);
2. *CLW and CCLW* to support continued reduction of manufacturing costs and to set the annual MCI targets at the level of each PFC process and manufacturing system level.

Since the relationship between productivity policy deployment (see Figure 3.1) and MCI policy deployment (see Figure 3.2) is made by the level required to be achieved by the annual MCI targets to meet the annual MCI goal to support internal manufacturing target profit and, indirectly, external manufacturing target profit, identifying the most credible and feasible CCLW at the level of each PFC process is the MCPD system key.

The pressure on KPIs targets in productivity policy deployment (see Figure 3.1) is set by the required level of losses and waste at KPIs level (losses and waste that are perceptible at process level). For example, to support the number of products sold by improving the product quality rate (OMI) and further, through the scrap ratio and by rework in the final stage ratio (level 1 KPIs), the level of MCI means targets for losses from equipment losses level (quality ratio) shall be established. In this example, quality issues are considered to be a credible assumption for each MCI strategy in order to establish an annual MCI target. In conclusion, Figure 3.2 sets the minimum level of KPIs targets in Figure 3.1. This minimum level must ensure the annual MCI goal or internal manufacturing profit ("I") (see top of Figure 3.1).

As has already been said, meeting the annual MCI goal and internal manufacturing profit ("I") will generate a self-driving effect on external manufacturing profit by increasing the number of products manufactured (maximizing outputs) and sold (minimizing inputs) based on a continuous and planned decrease of unit cost of manufacturing.

Once annual MCI targets have been established, based on actual opportunities for achieving the annual CCLW, annual MCI means are identified at each CLW level and, implicitly, at the level of each type of losses and waste from the processes that are affected by CCLWs of each PFC.

3.2.1.3 MCI Policy Deployment for Increase the Number of Products Sold

In order to really increase the number of products sold and to ensure an acceptable level of manufacturing profit, it is necessary to identify the losses and waste from the processes (see Figure 3.2) and then institute a real, continuous and consistent removal thereof. For this, there is a need for a more accurate localization and identification of losses and waste in the main processes of each PFC and the whole manufacturing system. Summing up the MCI policy deployment for each PFC provides an image for MCI policy deployment on the overall company (Figure 3.2).

Table 3.3 presents a first example of MCI policy deployment for a PFC (CPM: number of products sold; OMI: manufacturing capacity; KPIs for losses and waste: equipment losses; a section from Figure 3.2). This is the extension of the example in Table 3.1.

Losses are *time-related loss (TRL) and physical loss (PL)* for defects. Maintenance and quality managers are responsible for data collection (for scrap and rework). Data are collected by the teams in each area and the frequency of data collection is: (1) continuous (breakdown); (2) weekly (minor stoppages; tool changes); (3) monthly (setup, settings and adjustments; start-up; scrap and rework; and scheduling shutdown); and (4) quarterly (equipment cycle time). (Posteucă and Sakamoto, 2017, pp. 116–132).

KPIs levels for losses can be extended for more than level 4, such as for breakdown time of equipment (KPIs 4). KPIs 5 for breakdown time of equipment can be: time to failure of the tool; time with mechanical failures; time with electrical systems failures; time with hydraulic failures; time with soft failures, etc.

Table 3.4 presents a second example of MCI policy deployment for a PFC (CPM: number of products sold; OMI: delivery; KPIs for losses and waste: delivery waste; a section from Figure 3.2). It is the extension of the example in Table 3.2.

Therefore, first of all it is necessary to identify as accurately as possible the losses and waste at the level of each main process of each PFC and the whole manufacturing system.

TABLE 3.3

An Example of MCI Policy Deployment for a PFC for Equipment Losses (in Synthesis)

Company Productivity Vision through MCI	What do managers want?						Units	Years					
								"N-1"		"N" (Target)		"N+1" (Target)	
	CPM	OMIs	KPIs1	KPIs2	KPIs3	KPIs4		How much? fill in	Where? Processes/ System (*)	How much? fill in	Where? Processes/ System (*)	How much? fill in	Where? Processes/ System (*)
	Number of Products Sold - Increase						pieces	x	P/S (***)	x	P/S	x	P/S
		Manufacturing Capacity					pieces	x	P/S	x	P/S	x	P/S
			Equipment Losses (Current Strategic Equipments)				%	x	P/S	x	P/S	x	P/S
				Downtime Equipment Losses			%	x	P/S	x	P/S	x	P/S
					Breakdown Losses		min.	x	P/S	x	P/S	x	P/S
						Breakdown rate of equipment	min.	x	P/S	x	P/S	x	P/S
						Repair rate of equipment	min.	x	P/S	x	P/S	x	P/S
						Number of breakdowns in a period	no.	x	P/S	x	P/S	x	P/S
						Breakdowns time of equipment	min.	x	P/S	x	P/S	x	P/S
						Number of scrap produced by the breakdown	pieces	x	P/S	x	P/S	x	P/S

(Continued)

TABLE 3.3 (CONTINUED)

An Example of MCI Policy Deployment for a PFC for Equipment Losses (in Synthesis)

Company Productivity Vision through MCI	CPM	OMIs	KPIs1	KPIs2	KPIs3	KPIs4	Units	Years					
								"N-1"		"N" (Target)		"N+1" (Target)	
								How much? fill in	Where? Processes/ System (*)	How much? fill in	Where? Processes/ System (*)	How much? fill in	Where? Processes/ System (*)
						Number of rework produced by the breakdown	pieces	x	P/S	x	P/S	x	P/S
					Set-up, Settings, Adjustments Losses		min.	x	P/S	x	P/S	x	P/S
						Number of set-ups in a period	no.	x	P/S	x	P/S	x	P/S
						Set-up time of equipment	min.	x	P/S	x	P/S	x	P/S
						Number of scrap produced during the set-up	pieces	x	P/S	x	P/S	x	P/S
						Number of rework produced during the set-up	pieces	x	P/S	x	P/S	x	P/S
					Tool Changes Losses		min.	x	P/S	x	P/S	x	P/S
						Number of units produced between the two tool changes of equipment	pieces	x	P/S	x	P/S	x	P/S
						Tool changing time of equipment	min.	x	P/S	x	P/S	x	P/S

(Continued)

TABLE 3.3 (CONTINUED)

An Example of MCI Policy Deployment for a PFC for Equipment Losses (in Synthesis)

Company Productivity Vision through MCI	What do managers want?						Units	Years					
								"N-1"		"N" (Target)		"N+1" (Target)	
	CPM	OMIs	KPIs1	KPIs2	KPIs3	KPIs4		How much? fill in	Where? Processes/ System (*)	How much? fill in	Where? Processes/ System (*)	How much? fill in	Where? Processes/ System (*)
						Number of scrap produced during the tool changes	pieces	x	P/S	x	P/S	x	P/S
						Number of rework produced during the tool changes	pieces	x	P/S	x	P/S	x	P/S
					Equipment Start-up Time Losses		min.	x	P/S	x	P/S	x	P/S
						Number of start-ups in a period	pieces	x	P/S	x	P/S	x	P/S
						Start-up time of equipment	min.	x	P/S	x	P/S	x	P/S
						Number of defectives during the start-up time	pieces	x	P/S	x	P/S	x	P/S
				Performance Equipment Losses			%	x	P/S	x	P/S	x	P/S
					Equipment Cycle Time (Speed Down) Losses		min.	x	P/S	x	P/S	x	P/S
						Theoretical cycle time of equipment	sec.	x	P/S	x	P/S	x	P/S

(Continued)

TABLE 3.3 (CONTINUED)

An Example of MCI Policy Deployment for a PFC for Equipment Losses (in Synthesis)

Company Productivity Vision through MCI	What do managers want?						Units	Years					
								"N-1"		"N" (Target)		"N+1" (Target)	
	CPM	OMIs	KPIs1	KPIs2	KPIs3	KPIs4		How much? fill in	Where? Processes/ System (*)	How much? fill in	Where? Processes/ System (*)	How much? fill in	Where? Processes/ System (*)
						Actual cycle time of equipment	sec.	x	P/S	x	P/S	x	P/S
					Equipment Minor Stoppages Losses		min.	x	P/S	x	P/S	x	P/S
						Minor stoppages rate of equipment	min.	x	P/S	x	P/S	x	P/S
						Total repair time of equipment	min.	x	P/S	x	P/S	x	P/S
				Defect Equipment Losses			%	x	P/S	x	P/S	x	P/S
					Scrap and Rework Losses		min.	x	P/S	x	P/S	x	P/S
						Scrap rate of equipment /process	pieces	x	P/S	x	P/S	x	P/S
						Rework rate of equipment/process	pieces	x	P/S	x	P/S	x	P/S
				Temporal accepTable Equipment shutdown losses			%	x	P/S	x	P/S	x	P/S

(Continued)

TABLE 3.3 (CONTINUED)

An Example of MCI Policy Deployment for a PFC for Equipment Losses (in Synthesis)

Company Productivity Vision through MCI			*What do managers want?*				Units	Years					
								"N-1"		"N" (Target)		"N+1" (Target)	
	CPM	OMIs	KPIs1	KPIs2	KPIs3	KPIs4		How much? fill in	Where? Processes/System (*)	How much? fill in	Where? Processes/System (*)	How much? fill in	Where? Processes/System (*)
					Scheduling Shutdown Losses		min.	x	P/S	x	P/S	x	P/S
						Total time of downtime/month	min.	x	P/S	x	P/S	x	P/S

Note (*): processes: the main processes of each PFC; system: manufacturing system approach; (**): units of measurement are determined for all PFCs by setting an average of 5–15 representative products for each PFC; (***): P/S: Process/System (manufacturing system).

TABLE 3.4

An Example of MCI Policy Deployment for a PFC for Delivery Waste (in Synthesis)

Company Productivity Vision through MCI	CPM	OMIs	What do managers want?			Unit	Years					
			KPIs1	KPIs2	KPIs3		"N-1"		"N" (Target)		"N+1" (Target)	
							How much? fill in	Where? Processes/ System (*)	How much? fill in	Where? Processes/ System (*)	How much? fill in	Where? Processes/ System (*)
	Number of Products Sold - Increase					pieces	x	P/S (***)	x	P/S	x	P/S
		Delivery				days	x	P/S	x	P/S	x	P/S
			Supply Chain Lead Time			days	x	P/S	x	P/S	x	P/S
				Overall Material Inventory (waste)		pieces	x	P/S	x	P/S	x	P/S
					Raw materials inventory over maximum allowed	pieces	x	P/S	x	P/S	x	P/S
					Raw materials inventory under minimal allowed	pieces	x	P/S	x	P/S	x	P/S
					Components inventory over maximum allowed	pieces	x	P/S	x	P/S	x	P/S
					Components inventory under minimal allowed	pieces	x	P/S	x	P/S	x	P/S
					Production stoppages due to lack of materials/ components	min.	x	P/S	x	P/S	x	P/S
					Production stoppages due to the change in production planning	min.	x	P/S	x	P/S	x	P/S
					Supplier sources lost	days	x	P/S	x	P/S	x	P/S
			Manufacturing Lead Time			days	x	P/S	x	P/S	x	P/S

(Continued)

TABLE 3.4 (CONTINUED)

An example of MCI policy deployment for a PFC for delivery waste (in synthesis)

Company Productivity Vision through MCI	CPM	OMIs	What do managers want? KPIs1	KPIs2	KPIs3	Unit	"N-1" How much? fill in	"N-1" Where? Processes/System (*)	"N" (Target) How much fill in	"N" (Target) Where? Processes/System (*)	"N+1" (Target) How much fill in	"N+1" (Target) Where? Processes/System (*)
				WIP Inventory (waste)		pieces	x	P/S	x	P/S	x	P/S
					WIP Standard near equipment/lines over maximum allowed	pieces	x	P/S	x	P/S	x	P/S
					WIP Standard near equipment/lines under minimal allowed	pieces	x	P/S	x	P/S	x	P/S
					WIP from Set-up (on areas) over maximum allowed	pieces	x	P/S	x	P/S	x	P/S
					WIP from Set-up (on areas) under minimal allowed	pieces	x	P/S	x	P/S	x	P/S
					WIP from Transfer (on areas) over maximum allowed	pieces	x	P/S	x	P/S	x	P/S
					WIP from Transfer (on areas) under minimal allowed	pieces	x	P/S	x	P/S	x	P/S
				Total Cycle Time (waste)		sec.	x	P/S	x	P/S	x	P/S
					Total standard cycle time (minutes/piece) versus Total real cycle time (minutes/piece)	sec.	x	P/S	x	P/S	x	P/S
			Delivery Lead Time			days	x	P/S	x	P/S	x	P/S
				Finished Product Inventory (waste)		pieces	x	P/S	x	P/S	x	P/S

(Continued)

TABLE 3.4 (CONTINUED)

An example of MCI policy deployment for a PFC for delivery waste (in synthesis)

Company Productivity Vision through MCI	CPM	OMIs	KPIs1	KPIs2	KPIs3	Unit	"N-1"		"N" (Target)		"N+1" (Target)	
		What do managers want?					How much? fill in	Where? Processes/ System (*)	How much? fill in	Where? Processes/ System (*)	How much? fill in	Where? Processes/ System (*)
					Finished Product Stock over maximum allowed	pieces	x	P/S	x	P/S	x	P/S
					Finished Product Stock under minimal allowed	pieces	x	P/S	x	P/S	x	P/S
					Packaging Inventory over maximum allowed	pieces	x	P/S	x	P/S	x	P/S
					Packaging Inventory under minimal allowed	pieces	x	P/S	x	P/S	x	P/S
					Number of delayed trucks	no.	x	P/S	x	P/S	x	P/S
					Stand-by Finished products ratio	%	x	P/S	x	P/S	x	P/S

Note (*): processes: the main processes of each PFC; system: manufacturing system approach; (**): units of measurement are determined for all PFCs by setting an average of 5–15 representative products for each PFC; (***): P/S: Process/System (manufacturing system).

3.2.1.4 MCI Policy Deployment for Manufacturing Cost Decrease

In order to truly reduce the product unit cost of manufacturing and ensure an acceptable level of manufacturing profit, it is necessary to identify the cost of waste and waste (CLW) from the processes (see Figure 3.2), and then institute a real, continuous and consistent removal thereof (see Figure 3.2). For this, a more accurate localization and identification of CLW is needed at the level of the main processes of each PFC and the whole manufacturing system. Summing up CLW for all PFCs provides a picture of MCI policy deployment on the overall company (Figure 3.3).

Table 3.5 shows an example of total CLW for a PFC. By summing up all PFCs at company level, a total CLW is determined for the entire company. By determining the acceptable future status of CLW for the next five years (or the CLW level that can be approached), more precisely the TOCR, an annual productivity stake is determined through MCI Policy Deployment and, implicitly, the MCPD system.

In order to set targets for productivity policy and MCI policy deployment, a detailed analysis of the main processes of each PFC is needed to identify

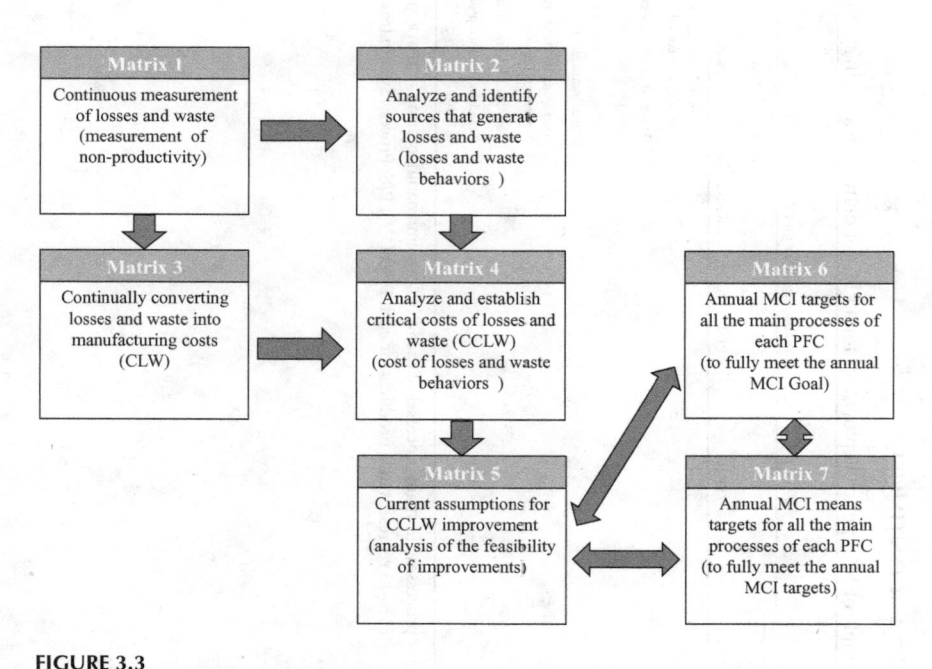

FIGURE 3.3

The seven matrices for establishing the annual MCI targets and means contribution to the annual MCI goal.

TABLE 3.5

MCI Policy Deployment for a CLW (in Synthesis for a PFC)

Company Productivity Vision through MCI	What do managers want?				Units (**)	Years					
						"N-1"		"N" forecasts		"N+1" forecasts	
	CPM	OMIs	KPIs1	KPIs2		How much? fill in	Where? Processes/ System (*)	How much? fill in	Where? Processes/ System (*)	How much? fill in	Where? Processes/ System (*)
	Manufacturing Cost				$	$	P/S (***)	$	P/S	$	P/S
		CLW			$	$	P/S	$	P/S	$	P/S
			CLW for Manufacturing Capacity		$	$	P/S	$	P/S	$	P/S
				Cost of Equipment Losses	$	$	P/S	$	P/S	$	P/S
				Cost of New Equipment Losses	$	$	P/S	$	P/S	$	P/S
				Cost of New Product Development Losses	$	$	P/S	$	P/S	$	P/S
			CLW for Manufacturing Control		$	$	P/S	$	P/S	$	P/S
				Cost of Human Work Losses	$	$	P/S	$	P/S	$	P/S
				Cost of Maintenance Material Losses	$	$	P/S	$	P/S	$	P/S
				Cost of Raw Material Losses	$	$	P/S	$	P/S	$	P/S
				Cost of Energy/Utilities Losses	$	$	P/S	$	P/S	$	P/S

(Continued)

TABLE 3.5 (CONTINUED)

MCI Policy Deployment for a CLW (in Synthesis for a PFC)

Company Productivity Vision through MCI	CPM	OMIs	KPIs1	KPIs2	Units (**)	"N-1" How much? fill in	"N-1" Where? Processes/System (*)	"N" forecasts How much? fill in	"N" forecasts Where? Processes/System (*)	"N+1" forecasts How much? fill in	"N+1" forecasts Where? Processes/System (*)
				Internal Logistics/Material Handling Losses	$	$	P/S	$	P/S	$	P/S
			CLW for Manufacturing Delivery		$	$	P/S	$	P/S	$	P/S
				Cost of Waste: Raw Material Inventory	$	$	P/S	$	P/S	$	P/S
				Cost of Waste: Components Inventory	$	$	P/S	$	P/S	$	P/S
				Cost of Waste: WIP from Set-up (WIP S)	$	$	P/S	$	P/S	$	P/S
				Cost of Waste: WIP from Transfer (WIP T)	$	$	P/S	$	P/S	$	P/S
				Cost of Waste: Finished Products Inventory	$	$	P/S	$	P/S	$	P/S
				Cost of Waste: Packaging Inventory	$	$	P/S	$	P/S	$	P/S

Note (*): processes: the main processes of each PFC; system: manufacturing system approach; (**): for each PFC by setting an average of 5–15 representative products for each PFC; (***): P/S: Process/System (manufacturing system).

opportunities for setting the annual MCI targets and means to meet the annual MCI goal.

Therefore, the purpose of this first step of the MCPD is to know the current state and the past context of policy productivity deployment and the MCI policy deployment in order to establish the annual MCI targets and means converging to the annual MCI goal (see Figure 3.1 above; KPIs 2: annual MCI targets to meet the annual MCI goal or annual MCI purpose). The main goal is to identify TOCR for the next five years. It also identifies the extent to which MCI targets have been met over time to identify the potential opportunities for the coming year. For each primary process of each PFC, besides KPIs specific to that process (KPIs of level 1, 2, ... "n"; linked to CPV; see Tables 3.1 and 3.2), the evolution in time and the current state are constantly monitored for: (1) CLW for each type of important product (for 5–15 products; the TOCR determination method); (2) the structure of variable and fixed costs affected by losses and waste (Pareto chart); and (3) the structure of losses and waste (Pareto chart). TORC is an important element of the company's strategic plan.

3.2.2 Step 2: Annual MCI Targets and Means

The second step of the MCPD system is MCI's most sensitive challenge: setting annual targets and means. For this, a continuous and consistent reconciliation between the top-down approach and the bottom-up approach is required to set the annual MCI target for each PFC by developing seven matrices:

- Matrix 1: continuous measurement of losses and waste;
- Matrix 2: analyze and identify sources that generate losses and waste;
- Matrix 3: continually converting losses and waste into manufacturing costs;
- Matrix 4: analyze and establish critical costs of losses and waste;
- Matrix 5: current assumptions for critical costs of losses and waste improvement;
- Matrix 6: setting annual MCI targets to achieve annual MCI goal;
- Matrix 7: setting annual MCI means to achieve annual MCI targets.

With the help of these seven matrices, it is possible to establish profitable projects of annual improvement that contribute to the full accomplishment of the annual MCI goal.

3.2.2.1 Reconciling Top-Down and Bottom-Up for Annual MCI Goal for Each PFC

In order to establish annual MCI targets amid means for each main process of each PFC and/or for each representative product of each PFC, a reconciliation needs to be made between the need to reduce the unit costs of products for the manufacturing stage (see Section 2.4 from the previous chapter) and the TOCR (Posteucă and Sakamoto, 2017, pp. 140–154).

This reconciliation is required at least once every six months, or whenever necessary to ensure that Principle No. 1 of the MCPD system is observed. In fact, the elements in Figure 3.2, those in Tables 3.3 and 3.4, and especially those in Table 3.5, are monitored at the level of each month of the current year (year "N"). Sometimes, to increase relevance and to analyze trends, if measurements exits, past events are evaluated for more than the previous year (e.g. three years ago).

If over a year, there are significant differences between what was planned and what was achieved at the level of achieving the annual MCI target, annual MCI means are re-planned. Annual MCI targets and means can be adjusted over a year, but the annual MCI goal does not change.

Through this reconciliation, it is sought to comply with Principle No. 4 of the MCPD system: *the continuous reconciliation of annual MCI targets and means for each PFC.* According to this principle, annual MCI targets are continually aligned to the annual MCI goal and, furthermore, each annual MCI means (*kaizen* and *kaikaku* for MCI) is aligned to the annual MCI target.

Through the mechanism of reconciliation of the annual MCI targets and means to the annual MCI goal, materialized in Figure 3.2, starting from the current state of the TOCR, the ideal status of the CLW is set for the next five years and the future state for next year. Reconciliation is achieved in accordance with all the basic principles of the MCPD system (see Figure 1.10) (Posteucă, 2018, p. 167–173). The final outcome of the reconciliation process is the establishment of current assumptions for identifying the CCLW: credible assumptions for each MCI strategy (CD1 to CD8); uncertain assumptions for each MCI strategy (CD1 to CD8) and vulnerable assumptions for each MCI strategy (CD1 to CD8) (see Table 2.1 and Section 2.4 of the previous chapter). With the help of the current assumptions to identify CCLW, annual MCI targets and means are established. Figure 3.3 shows the basic logic of the seven matrices that aim to plan the harmonious transformation of the manufacturing flow of each PFC, as follows:

- Matrix 1: It evaluates the current level of non-productivity by continually measuring losses and waste in all the main processes of each PFC and manufacturing system, namely by measuring and monitoring losses and waste KPIs (see Figure 3.2 and Table 3.3 and 3.4) (Posteucă and Sakamoto, 2017, pp. 119–122); Matrix 1 is a qualitative analysis;
- Matrix 2: It analyzes losses and waste behaviors to identify the root causes that generate losses and waste across major processes for each PFC and at manufacturing system level;
- Matrix 3: It aims to continuously convert losses and waste into manufacturing costs (CLW) for all the main processes of each PFC and at manufacturing system level to identify on a continuous basis the part of the manufacturing cost structure that is affected by CLW or costs behind losses and waste (see Table 3.5); Matrix 3 is a quantitative analysis;
- Matrix 4: It determines the cost of waste and waste behaviors to identify losses and waste and roots the causes along all the processes of a PFC and manufacturing system level;
- Matrix 5: This is the matrix that analyzes current assumptions with the aim to identify the most credible and feasible CCLWs that can achieve the annual MCI targets and, implicitly, the annual MCI goal by choosing the most pertinent annual MCI means targets. This is the point where the actual process of reconciliation based on the MCI catchball process at all hierarchical levels in the company is achieved to ensure that each annual PFC is convergent and contributes to consistent achievement of CPV and CPM;
- Matrix 6: This is based on Matrix 5 and establishes the acceptable level of annual MCI targets that will ensure the full achievement of the annual MCI goal;
- Matrix 7: This is the matrix that sets the most feasible annual MCI means to meet the annual MCI targets (KKI). Matrix 7 aims to set targets for losses and waste in Matrix 1 to ensure a harmonious transformation of manufacturing flow across all PFCs and at the level of the manufacturing system and, implicitly, MCI policy deployment for a PFC (see Figure 3.2) and productivity policy deployment (see Figure 3.1).

In this context, the following sections will outline in more detail the seven matrices that have the ultimate goal of scientifically establishing the contribution of each annual MCI targets and means of each PFC (see Figure 3.3) to meet the annual MCI target by planning the harmonious transformation of the manufacturing flow of each PFC and the planning of harmonious transformation at the manufacturing system level.

Therefore, each manufacturing process of a PFC will have outputs (products, services and information) affected by: (1) the level of losses and waste in that process; (2) the level of losses and waste of the manufacturing system; or (3) the level of losses and waste that affect the process from other upstream processes and/or downstream manufacturing flow.

Some of the inputs in a manufacturing process will be found in finished products, and some of those inputs will be regrouped to losses and waste and CLW. The manufacturing flow creating added-value inputs perceived by customers is often well-defined and tracked. Instead, the part of the manufacturing flow that determines the occurrence of losses and waste and associated CLWs is less well-known within companies, especially in view of the changing behavior of losses and waste over time and their impact on the level of manufacturing target profit. Sometimes it is considered that the current level of manufacturing profit, largely made up of external manufacturing profit, gained on sales, is high enough to justify the effort to identify and consistently address CLW.

3.2.2.2 Matrix 1: Continuous Measurement of Losses and Waste

In order to identify the opportunities for a consistent approach to MCI, a qualitative analysis of the losses and waste levels for each main process of each PFC and manufacturing system is needed, with the help of KPIs for losses and waste; in fact, the level of non-productivity is measured on a continuous basis (in the last 6–12 months).

Once the annual MCI goal is established, the directions to reach it are sought, especially at the level of the main processes of each PFC. The identification of the main processes is determined by analyzing the technological flow diagrams of the parts of each PFC: technological phases; technological processes; technological operations; number of operations; name of operations.

For example, a company in automotive components industry may have:

- *Technological phases:* injection of plastic parts (I) and painting of plastic parts (II);
- *Technological processes* (for injection of plastic parts I): (1) drying granules; (2) preparation of the injection process; (3) injection

process; (4) download the injected parts; (5) packing; (6) storage; and (7) transport;

- *Technological operations* (for (1) drying granules: material feed–(A) granules; (B) opening of sacks with granules; (C) feeding of granular bunkers; (D) setting drying temperatures; (E) self-inspection;
- *Number of operations* ((A) for material feed: 1, 2, 3, 4, ...;
- *Name of operations* ((A) for material feed: a, b, c, d, ...

In order to facilitate the measurement of losses and waste, the type of equipment used and the process parameters of the equipment is specified for each technological process (e.g. for (1) drying granules, the following is known: the equipment–"AP" drying and feeding; the process parameters; the number of operators; the description of the work tasks and the type of raw materials used).

This detailed manufacturing flow structure for each main process of each PFC is required to evaluate all losses and waste directly at the source where the events take place. Continuous measurement should provide the most accurate picture of losses and waste for at least 12 months. At the same time, the continuous monitoring of these losses and waste levels checks must be carried out: (1) the current level of losses and waste compared to the level that was planned to be achieved on the basis of *kaizen* and *kaikaku* implementations (previous MCI means); (2) losses and waste trends; and (3) the degree of contribution of the improvement of losses and waste for achieving the CPV and CPM (required capacities compared to current ones).

The basic questions for which answers are sought with the help of Matrix 1 are:

- What activity/technological operations generate losses and waste?
- In what process/technological processes is localized n losses and waste generating activity?
- What types of losses and waste occur in each process?
- How does the system generate losses and waste?
- When is a certain type of lost or waste activated?
- How are losses and waste measured?
- How often and how much does each type of losses and waste occur?
- What trend do losses and waste have in each process?
- How is the trend of losses and waste evaluated?
- Who collects losses and waste and at what time intervals?

Figure 3.4 shows the basic structure of Matrix 1.

As one can see, the loss and waste valuation is performed for the entire technological flow of a PFC ("PFC 1"). Horizontally, there are the 14 types of losses and waste (LW), and vertically there is the detail of PFC 1 (technological phases, technological processes and technological operations).

Each process and each PFC is evaluated from the perspective of the 14 LW categories at the level of:

- Losses and waste at the level of each process operation (*total points per operation*–vertically; the penultimate column);
- Losses and waste at process level (*process/LW*) and at PFC level (*total "PFC 1"/LW*);
- The manufacturing system influences on LW in the process reviewed (*system/LW*) and total PFC (*total "PFC 1" system/LW*);
- SDI for each process (*process/SDI*) and total PFC (*Total "PFC 1"/SDI*);

The basic purpose of Matrix 1 is to achieve a first targeting of the most important losses and waste. Their processes and operations with the highest score will be those that will be investigated first. In this way, a method of warning is established on all losses and waste for each PFC.

The impact on the process is assessed from the point of view of:

1. The frequency of occurrence of losses and waste;
2. The duration of an event, and
3. The associated inventory level (such as the structure of how KPIs are created for LW; see Table 3.3–KPIs level 4; see Table 3.4–KPIs level 3).

The impact of LW is on the method of achieving the *product number* for:

- *Manufacturing capacity (SDI 1)*: especially *time-related loss (TRL)*;
  - Continuous measurement is carried out at the level of each main process of equipment–based PFC: LW1-3;
  - Monthly assessment is performed according to the frequency and duration of the occurrence of losses and according to the associated inventory/WIP level;
  - For e.g. *set-up, settings, adjustment time losses* for a process, it is considered to have a great impact on the process by determining the frequency of occurrence, such as 25 times per month, by

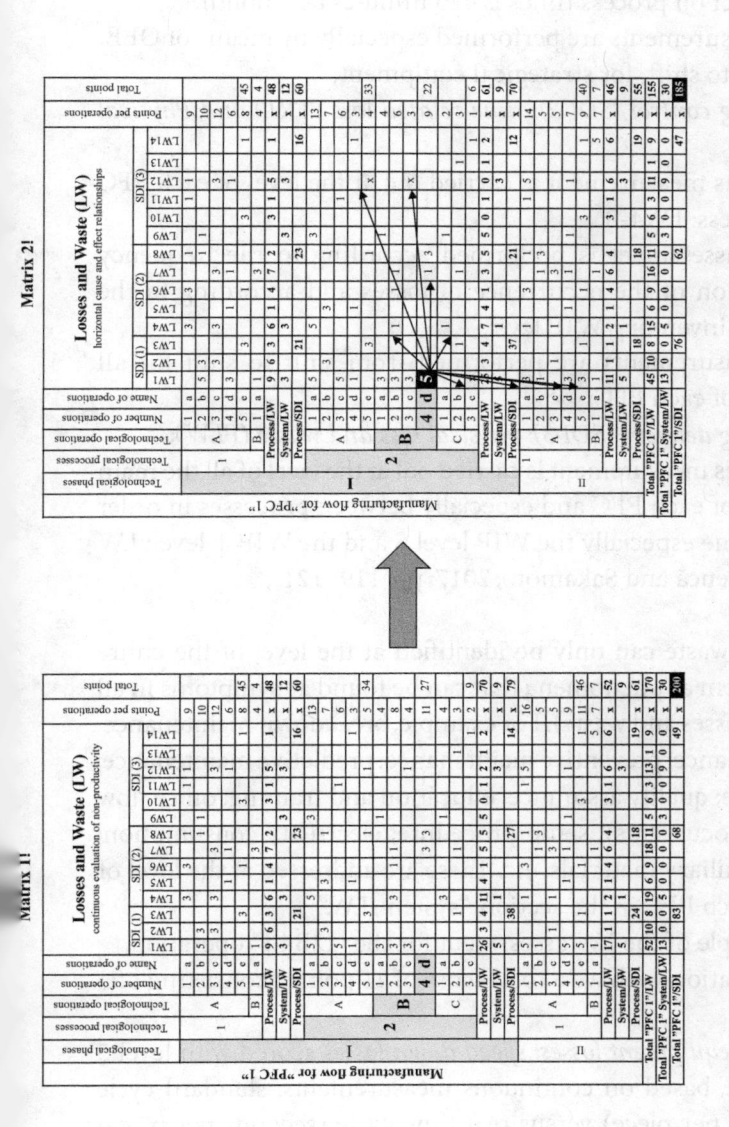

FIGURE 3.4

Matrix 1 for assessing losses and waste and Matrix 2 for identifying and evaluating the losses and waste behaviors along the main processes of a PFC.

Legend: SDI–Strategic direction of improvement (1): Manufacturing Capacity (Maximizing Outputs "E"); SDI–Strategic direction of improvement (2): Manufacturing Control (Minimizing Inputs "I"); SDI–Strategic direction of improvement (3): Manufacturing Delivery (Minimizing Inputs "r");

LW1: Equipment Losses (%, hours); LW2: New Product Development Losses (%, hours and $) –effects of product development over losses and waste in processes; LW3: New Equipment Development Losses (%, hours and $) –effects of product development over losses and waste in processes; LW4: Human Work Losses (%, hours) –fixed costs; LW5: Maintenance Material Losses (%, $) –fixed costs; LW6: Raw Material Losses (%, $) –variable costs; LW7: Energy/Utilities Losses (%, $) –variable costs; LW8: Internal Logistics/Material Handling Losses (distance; $) –variable costs; LW9: Waste: Raw Materials Inventory (number); LW10: Waste: Components Inventory–(if they are); LW11: Waste: WIP Set-up (WIP S) (minutes or units); LW12: Waste: WIP Transfer (WIP T) (minutes or units); LW13: Waste: Finished Products Stock (number); LW14: Waste: Packaging Inventory (number).

1 point–low impact due to the process. 3 points – average impact due to the process. 5 points–high impact due to the process.

determining the average duration of the event, such as 17 minutes, and by determining the WIP resulting from the set-up event, settings, adjustment time losses (WIP S), such as 386 pieces. The total impact on process times is 425 minutes per month.
- These measurements are performed especially by means of OEE, from shift to shift, for strategical equipment;
- *Manufacturing control (SDI 2): time-related loss (TRL) and Physical Loss (PL);*
  - Continuous measurement is carried out at the level of each PFC main process: LW4-8;
  - Monthly assessment is performed according to the frequency and duration of the occurrence of losses and according to the associated inventory/WIP level;
  - These measurements are performed from shift to shift for all processes of each PFC;
- *Manufacturing delivery (SDI 3): physical loss and waste (PLW);*
  - Continuous measurement is carried out at the level of all the main processes of each PFC and especially between processes in order to determine especially the WIP level S and the WIP T level: LW 9–14 (Posteucă and Sakamoto, 2017, pp. 119–121);

Some losses and waste can only be identified at the level of the entire manufacturing system as phenomena that can be found as symptoms in all processes (system losses and waste). For example, breakdown maintenance; time-based maintenance; preventive maintenance; predictive maintenance; quality maintenance; quality assurance; education and training; office flow (information and documents); setup procedure; electricity consumption; consumption of auxiliary materials, etc. These are addressed at the level of each process and each PFC in the section: "system/LW."

Below is an example of the LW assessment for the "I2B4d" operation.

For "I2B4d" operation we have a total score of 11 points, consisting of:

- LW1 is about *equipment losses; speed down losses* scored with level 5 of importance, based on continuous measurements: standard cycle time (seconds per piece) versus real-time cycle (seconds per piece) (see Table 3.3),
- LW7 is about *energy equipment losses* scored with level 3 of importance, based on continuous consumption measurements: theoretical consumption versus current consumption for "1A3b";

- LW8 is about *internal logistics losses* scored with level 3 of importance, based on continuous measurements of the internal logistics lead time of each route traveled to supply workstations with raw materials and/or components: theoretical and actual distances for "1A3b."

Furthermore, the following assessments were performed at the level of *technological processes "I2"*: "Process/LW"–70 points; "System/LW"–9 points and summed up "Process/SDI"–79 points. The biggest improvement opportunities in Matrix 1 for *technological processes "I2"* were for LW1 and LW4.

The total score for PFC 1 was 200 points. This score must decrease by the improvements decided in Matrix 7. To start targeting MCI enhancements, Matrix 2 will be drawn (continue the example for Section "I2B4d").

3.2.2.3 Matrix 2: Analyze and Identify Sources that Generate Losses and Waste

Throughout the processes of a PFC manufacturing flow, the occurrence and manifestation of a loss or waste will have effects both in the process of the event, and especially in the processes in the upstream and the downstream of the manufacturing flow. Continuous and early identification of losses and waste that cause other losses and waste along the manufacturing flow is the key to addressing consistent improvements in capacity increase (LW1, LW2 and LW3), rigorous manufacturing control increase (LW4–LW8) and continued delivery time decrease (LW9–LW14). Moreover, unplanned changes to manufacturing capacity and manufacturing control will generate effects on manufacturing delivery, or in other words, "waste (stocks) elasticity on losses" (Principle No. 7 of the MCPD system; see Figure 1.10) (Posteucă and Sakamoto, 2017, pp. 85–86; Posteucă, 2018, pp. 13–15; Posteucă, 2015, pp. 124–125).

To address the causes of waste and waste at their root where they are formed and to prevent them from occurring is more efficient and more effective than fighting their scattered effects throughout the entire manufacturing flow and at the level of the entire manufacturing system, effects most often found in inventory and stocks.

The basic questions for which answers are sought with the help of Matrix 2 are:

- What is the transversal behavior of losses and waste over time?
- What is the causal relationship of losses and waste across the entire manufacturing flow for a PFC and beyond?
- Which losses/waste influence which other losses/waste?
- Where are real changes in processes and in the system necessary to mitigate impact losses and waste on the number of products manufactured and delivered?

Continuing the example in Matrix 1 in the "I2B4d" section on LW1 (*equipment losses: speed down losses*), scored with level 5 of importance, Figure 3.5 shows the cause–effect analysis for *speed down losses* across the processes of a PFC, to see the impact of speed down losses on the other categories of losses and waste for upstream and downstream processes and on the "I2" process (the process where the equipment speed down losses events take place) in terms of capabilities, control and PFC manufacturing flow deliveries.

Based on the cause–effect analysis of losses and waste and Matrix 2, the *losses and waste stratification flow analysis (LWSFA)* is determined; this is the total measurement of the impact of equipment speed down losses on all affected losses and waste. It seeks to identify as many possible effects as possible. For example, a five-second deviation from standard-time equipment can have measurable effects on losses and waste for:

- *Upstream processes:* Increasing equipment idling for downstream processes with 1,620 seconds per month (LW1);
- *Downstream processes:* Increasing raw materials inventory (unplanned increase of WIP T) by 0.5% per month (LW9);
- *Process in which the equipment speed down losses event occurs:* WIP T increase (unplanned increase) for downstream processes by 1.2% per month, etc.

Without any preconceived ideas in the subsequent LW causal approach, with *kaizen* and *kaikaku* projects for MCI (Matrix 7), looking for and analyzing actual facts and facts behind LW, the past potential causes of the current outbreak of *equipment speed down losses* for the analyzed process, the following can be identified:

- LW1: a defective maintenance or improper environment for equipment (or environment that has become inappropriate) such as: cleanliness; brightness; vibration; temperature; humidity;

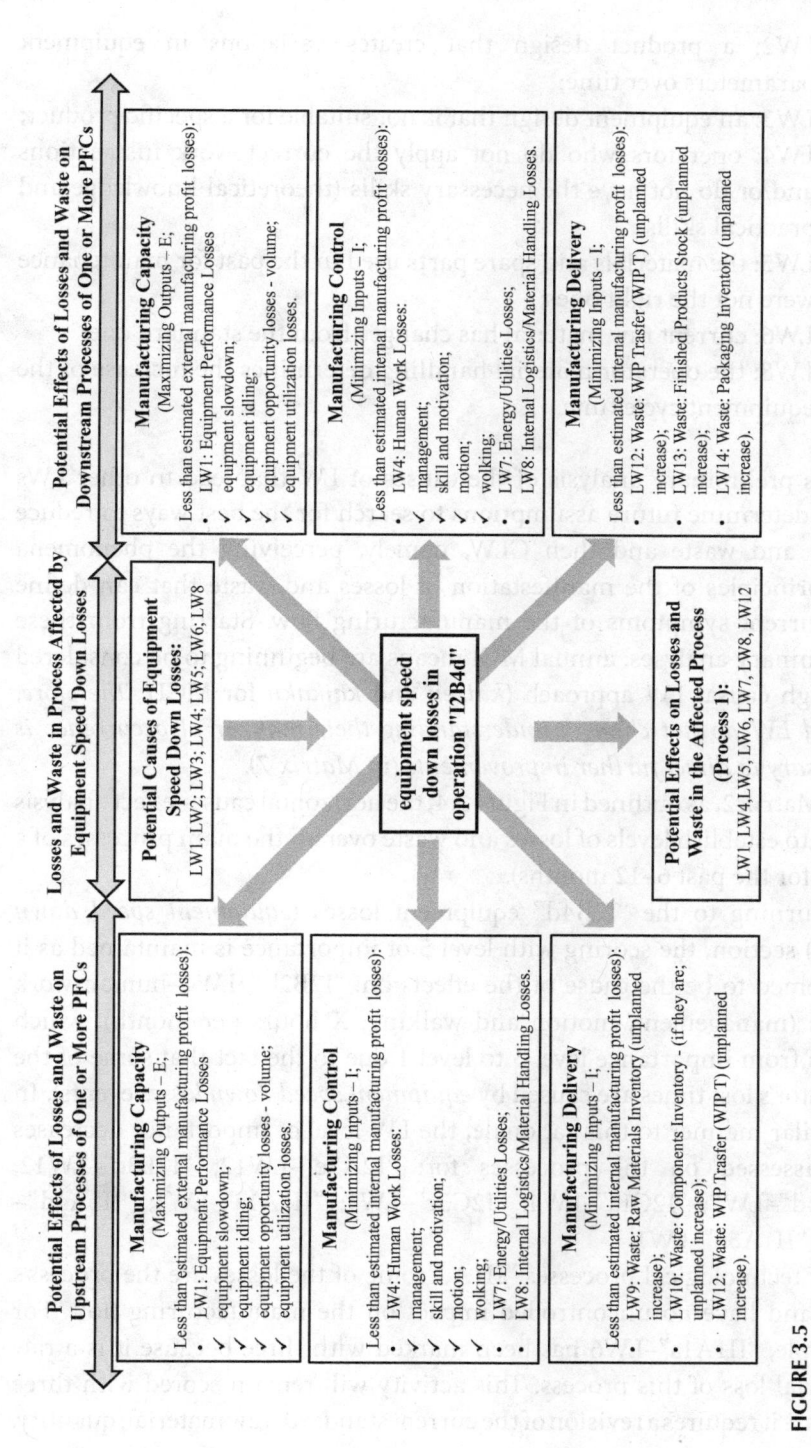

FIGURE 3.5

Example of cause–effect analysis in in operation "12B4d" along PFC processes for equipment speed down losses behavior.

- LW2: a product design that creates variations in equipment parameters over time;
- LW3: an equipment design that is not suitable for a specific product;
- LW4: operators who do not apply the correct work instructions and/or do not have the necessary skills (theoretical knowledge and practical skills);
- LW5: the materials and spare parts used in the past for maintenance were not the right ones;
- LW6: current raw material has changes from the standard one;
- LW8: the operator material handling determines the increase of the equipment cycle time.

This preliminary analysis of the causes of LW that lead to other LWs helps determine future assumptions to search for the best ways to reduce losses and waste and then CLW, namely, perceiving the phenomena and principles of the manifestation of losses and waste that can define the current symptoms of the manufacturing flow. Starting from these preliminary analyses, annual MCI means are beginning to be considered through causal LW approach (*kaizen* and *kaikaku* for MCI). *Therefore, causal LW do not change, understanding their manner of occurrence is necessary to direct further improvements (in Matrix 7).*

In Matrix 2, as outlined in Figure 3.4, the horizontal cause–effect analysis seeks to establish levels of losses and waste over all the main processes of a PFC (for the past 6–12 months).

Returning to the "I2B4d" equipment losses (*equipment speed down losses*) section, the scoring with level 5 of importance is maintained as it is deemed to be the cause of the effects on: "I2B2b", LW4–human work losses (management, motion and walking; X hours per month), which drops from importance level 3 to level 1 due to the fact that some of the operator's lost times are caused by *equipment speed down losses* events). In a similar manner to this rationale, the LW level of importance decreases are assessed on the processes for: "I2A5e"–LW12; "I2B3c"–LW12; "I2B4d"–LW7; "I2C1a"–LW4; "I2C3c"–LW1; "II1A1a"–LW1; "II1A4d"–LW1; "II1A5e"–LW1.

For technological processes "II1A," some of the losses are the process's own and have no uncontrolled impact on the manufacturing flow. For example, "II1A1a"–LW6 has been marked with three because it is a raw material loss of this process. This activity will remain scored with three because it requires a revision of the current standard (raw material: quantity, unit costs and conversion costs; for this, a series of discussions will be initiated–Material Loss Analysis (MLA)).

Some LWs in Matrix 2 that were scored with five points in Matrix 1 (five points: high impact due to the process) were scored with three points after the LW path impact analysis that affected that section. It is the cases of "II1A4d" and "II1A5e" that are affected by "I2B4d" equipment losses (*equipment speed down losses*). Similarly, other LWs can be downgraded in Matrix 2 against Matrix 1. The final result of Matrix 2 is the decrease of the total score from 200 points to Matrix 1 to 185 points, only from "I2B4d" (causal losses). In Matrix 2 there are still six operations scored with five points. These, along with other operations scored with three points must provide the answer to about 80% of the LW for PFC 1 (aiming to abide by the *Pareto principle*). In this way, annual MCI targeting begins to shape into a clearer outline.

In the shift meetings (daily MCI management), each process/equipment shutdown time and each type of inventory and stocks (waste) are analyzed and the causal relationship between waiting times and inventory and stocks (waste; in particular WIP S) are established. This information of the horizontal causality between losses and waste is information written in the shift report by the shift manager. At the end of the month, they know exactly how to settle the non-productivity of equipment and processes and their causal relationships, respectively:

- *Gemba* (actual place): the workplace, the place where losses and waste creation events occur and where they are visible;
- *Gembutsu* (actual things): current losses and losses that are affected by losses and considered root causes;
- *Genjitsu* (actual facts): the phenomenon that causes the occurrence of losses and considered root causes;
- *Genri* (principles): the theoretical and practical principles that determine the occurrence of losses and waste considered to be the root causes;
- *Gensoku* (standards and parameters): current standards and operational parameters (standard operating procedure (SOP)) that facilitate the emergence of losses and waste considered to be root causes.

Therefore, it was intended to observe the *Pareto principle* horizontally, transversely, along the entire manufacturing flow of PFC (*roughly 80% of the effects come from 20% of the causes*) to determine causal LW. The 100% justification of how LW occurs along the manufacturing flow of a PFC is the aim. In this way, *the LW causal accumulation pools* are identified at the level of the entire manufacturing flow of a PFC, pools that attract the resulting LW, both from the process of the event of losses or waste, as well as from previous and subsequent processes.

Therefore, in Matrix 2, causal LWs are LW caused by a problem of a process or equipment, an unplanned shutdown (shorter or longer). Emerging LWs are those that result from failure, slowing the speed of an equipment, or stopping or slowing down a process (materials, workforce, energy, etc.).

For example, for equipment, usually, equipment idling losses (LW1–equipment performance losses) or waiting for raw materials/components (LW1–independent equipment shutdown conditions not caused by equipment) are resultant LW and not causal LW. Other examples of LW results for equipment are: allowable time for cleaning equipment; allowable time for checking and lubricating equipment, allowable time for planned maintenance of equipment and time to power outages (no electricity).

Examples of causal (unplanned) problems losses for equipment are shown in Figure 3.6, while examples of causal losses for human work are presented in Figure 3.7.

Similarly, each manufacturing company sets its own causal LW at the level of the equipment and processes of each PFC depending on the specifics of the business. The basic work tool used to identify causal LW is *why–why analysis.*

The experience of managers and operators and LW measurement being as accurate as possible in Matrix 1 and when the LW events took place, are decisive in choosing causal LW at the level of equipment and, especially, the processes.

In order to target MCI improvements through CLW, Matrix 3 and Matrix 4 will be drawn (continued example for operation "I2B4d").

3.2.2.4 Matrix 3: Continually Converting Losses and Waste into Manufacturing Costs

As outlined in Section 2.3, in order to achieve a continuous decrease of product unit cost of manufacturing and to ensure a competitive price level, MCI core strategy aims at *accurately identifying the current status of the CLW in the main processes of each PFC.* The assessment in Matrix 3 is based

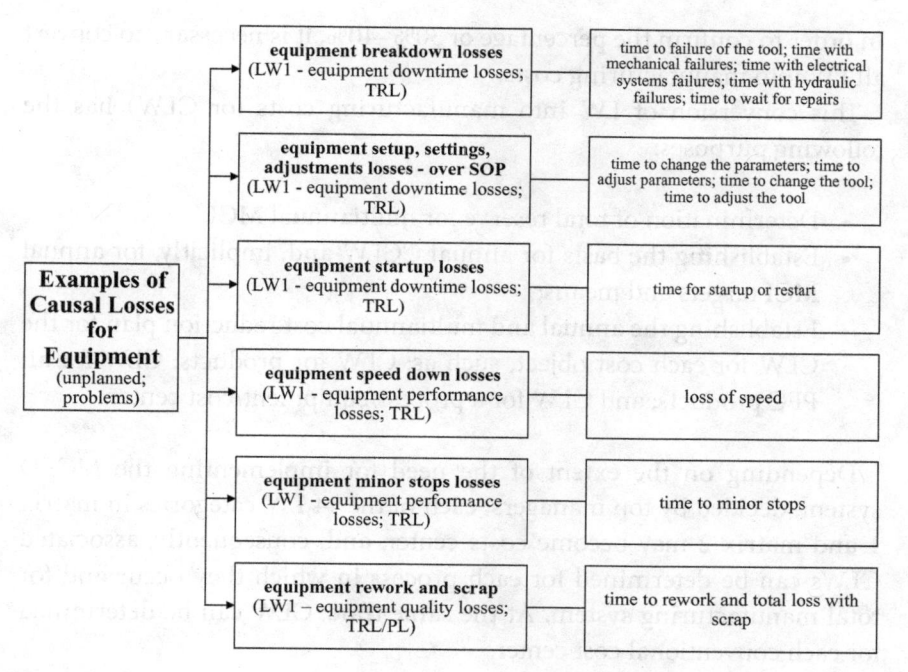

FIGURE 3.6
Examples of causal losses for equipment.

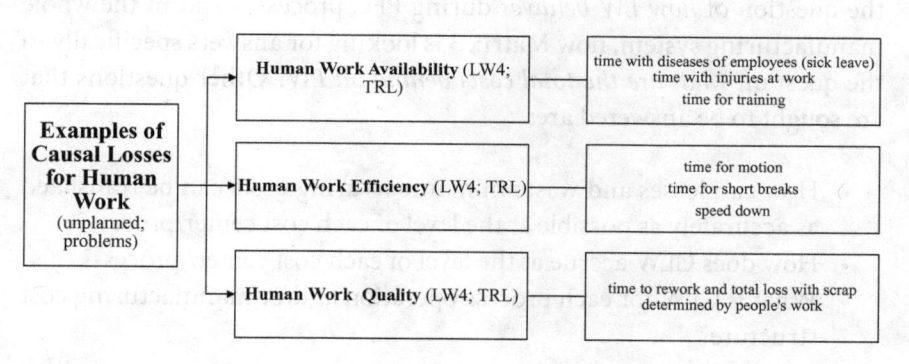

FIGURE 3.7
Examples of causal losses for human work.

on a quantitative analysis because most LW in Matrix 1 scored with the same score have different CLW. Usually, CLW is identified in a proportion of 30%–40% of the actual manufacturing costs. This is the assessment of the effectiveness of the work in Matrix 1. If the 30%–40% of the actual manufacturing costs for a PFC or the whole manufacturing company are not reached, then the rigor of identifying the 14 LW categories is increased.

In order to confirm the percentage of 30%–40%, it is necessary to convert all LW into manufacturing costs.

This conversion of LW into manufacturing costs (or CLW) has the following purposes:

- Determination of total reserve for multiannual MCI;
- Establishing the basis for annual CCLW and, implicitly, for annual MCI targets and means;
- Establishing the annual and multiannual cost reduction plan for the CLW for each cost object, such as: CLW for products; CLW for all PFC products; and CLW for a process/equipment/cost center.

Depending on the extent of the need for implementing the MCPD system, decided by top managers, each of the 14 LW categories in matrix 1 and matrix 2 may become costs center, and, consequently, associated CLWs can be determined for each process in which they occur and for total manufacturing system. At the same time, CLW can be determined for each conventional cost center.

While Matrix 1 searched for the answers to the question of *where is LW?*, in process or system, and while Matrix 2 searched for the answers to the question of *how LW behaves* during PFC processes and in the whole manufacturing system, now Matrix 3 is looking for answers specifically to the question *what are the total costs behind all LW?* Other questions that are sought to be answered are:

- How can losses and waste in manufacturing costs can be translated as accurately as possible at the level of each cost center/process?
- How does CLW accrue at the level of each cost center/ process?
- What is CLW for each process operation and/or manufacturing cost structure?

Once LW is evaluated through Matrix 1 and Matrix 2 to highlight how and where LW and causality relationships occur on the horizontal manufacturing flow of each PFC, it is time to face manufacturing costs with LW (Matrix 3). For this, it is first necessary to know the fixed costs, and especially the variable costs at the cost centers, usually the main processes of each PFC.

Typically, in practice, manufacturing companies use two large types of costing system:

1. *Normal costing system*: uses current costs for direct materials and direct labor costs but normal costs for "factory overhead." *Normal costing* involves estimating the allocation of part of the "factory overhead" for each product. Thus, an estimate of costs can be provided in a timely manner. At the same time, the normal costing system implies choosing an *appropriate level of activity* for allocating fixed overhead costs.

2. *Standard costing system*: uses standard costs for all cost elements (direct and indirect). Standard costs are the costs that manufacturing companies should achieve under efficient operating conditions. The standard costing system provides a basis for cost control, performance appraisal and process improvement.

Companies are less likely to use the *actual costing system* or use actual costs as the cost of the product, as unit costs fluctuate significantly, thus increasing the possibility of error in terms of prices. Moreover, there are difficulties in assessing performance. Since factory overhead costs are known only at the end of the period, there are few opportunities to reduce them and little information available when designing customer offers.

At the same time, some manufacturing companies use *activity-based costing* to more accurately allocate overhead costs, focusing on company activities by using multiple overhead cost bases that rate overhead cost direct labor hours as the cost driver or using machine hours as the cost driver, just as for normal and standard costing system.

From the perspective of the MCPD system, the manufacturing costs structure is as follows:

- Transformation costs
  - Direct labor costs (variable or fixed costs, depends on the company organization)
  - Indirect labor costs (variable costs)
  - Manufacturing overhead (variable costs, in general)
    - Maintenance costs/spare parts cost (variable or fixed costs, depends on the company organization)
    - Utility costs (variable costs)
    - Tool costs (variable costs)
    - Die and jig costs (variable costs)
  - Depreciation costs (fixed costs)
- Material costs (variable costs)

- Direct material costs
- Indirect material costs (auxiliary materials cost) (Posteucă, 2018, p. 93)

For example, for the allocation of manufacturing costs to the most common method of costing in manufacturing companies, normal costing (actual cost of materials, actual cost of labor and predetermined overhead rates), three methods of allocating the manufacturing cost (fixed costs and variable and direct and indirect costs) are used on cost centers or departments (the first level of cost objects):

- *Allocation on base of real data*: the allocation basis was man-hours by product for direct labor costs (regular and part-timer employee; variable cost), and the allocation basis for direct material costs (variable cost) is the current consumption per unit;
- *Allocation on the basis of monthly data* (as accurate as possible): energy (consumption-based allocation; electricity and fuel; variable cost), depreciation (allocation based on operating time; straight-line method; fixed costs),
- *Apportionment on base of monthly data*: the allocation basis is determined for maintenance/spare parts costs (fixed/variable costs), tool costs (variable costs), die and jig costs (variable costs), indirect labor costs (fixed/variable costs) and indirect material costs/auxiliary material costs (fixed/variable costs).

After these allocations of manufacturing cost per cost centers or departments (first level of cost objects), allocation is then made to the next level of cost objects (level 2) at the level of products and services.

Since the MCPD system is not a costing system, but a systematic and systemic improvement system of manufacturing costs, it is necessary to determine the LW associated part in the manufacturing cost structure. It is usually intended to have an acceptable accuracy of CLW for the last 6–12 months for each main process of each PFC and for the overall company.

Figure 3.8 shows an example of a detailed analysis of the CLW structure for a major PFC process (for one month). In this example, this PFC process was considered a cost center, the manufacturing speed was given by equipment, and there were no relevant LW records for LW2 and LW3 (Posteucă, 2018, p. 152; Posteucă and Sakamoto, 2017, p. 151 and pp. 193–193).

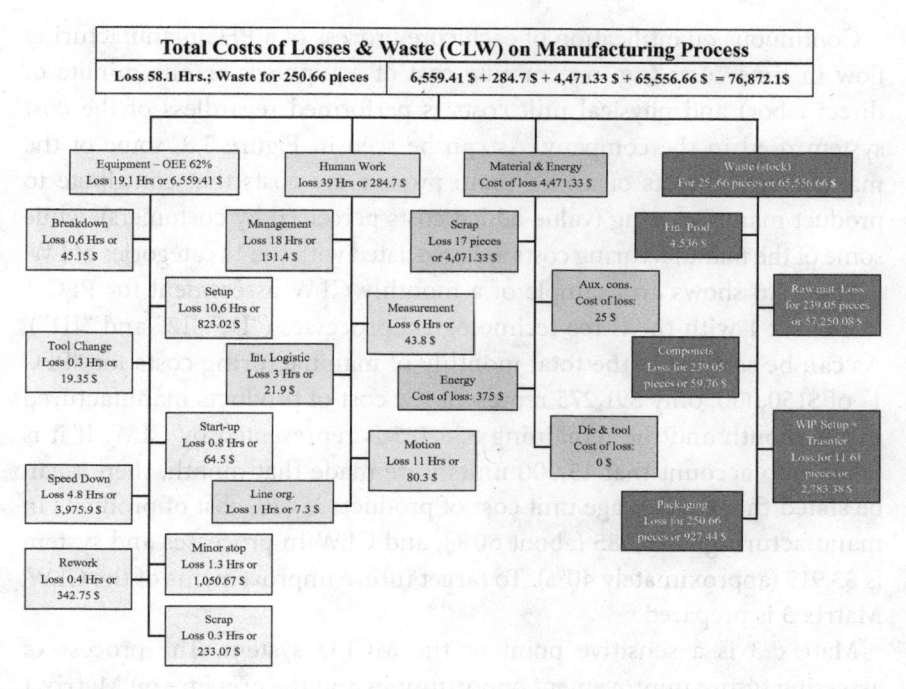

Total Costs of Losses & Waste (CLW) on Manufacturing Process

Loss 58.1 Hrs.; Waste for 250.66 pieces | 6,559.41 $ + 284.7 $ + 4,471.33 $ + 65,556.66 $ = 76,872.1$

Equipment – OEE 62%
Loss 19,1 Hrs or 6,559.41 $

Human Work
loss 39 Hrs or 284.7 $

Material & Energy
Cost of loss 4,471.33 $

Waste (stock)
For 25,66 pieces or 65,556.66 $

Breakdown
Loss 0,6 Hrs or
45.15 $

Management
Loss 18 Hrs or
131.4 $

Scrap
Loss 17 pieces
or 4,071.33 $

Fin. Prod.
4.536 $

Setup
Loss 10,6 Hrs or
823.02 $

Measurement
Loss 6 Hrs or
43.8 $

Aux. cons.
Cost of loss:
25 $

Raw mat. Loss
for 239.05 pieces
or 57,250.08 $

Tool Change
Loss 0.3 Hrs or
19.35 $

Int. Logistic
Loss 3 Hrs or
21.9 $

Energy
Cost of loss: 375 $

Componets
Loss for 239.05
pieces or 59.76 $

Start-up
Loss 0.8 Hrs or
64.5 $

Motion
Loss 11 Hrs or
80.3 $

Die & tool
Cost of loss:
0 $

WIP Setup +
Transfer
Loss for 11.61
pieces or
2,783.38 $

Speed Down
Loss 4.8 Hrs or
3,975.9 $

Line org.
Loss 1 Hrs or 7.3 $

Packaging
Loss for 250.66
pieces or 927.44 $

Rework
Loss 0.4 Hrs or
342.75 $

Minor stop
Loss 1.3 Hrs or
1,050.67 $

Scrap
Loss 0.3 Hrs or
233.07 $

FIGURE 3.8
The CLW structure for a main manufacturing process for a PFC (cost center for a month).

Therefore, if the allocation of variable and fixed costs per cost centers (departments or physical areas) and then the allocation of direct and indirect costs to products and services is a common practice in manufacturing companies, the allocation of variable and fixed costs and then the direct and indirect costs related to the 14 categories of LW is not a common practice in many manufacturing companies (there is no cost accounting for the 14 LW categories for each accounting period and for each month separately). Moreover, by assessing the conventional performance of cost structures, such as investigating the variation between the *standard cost* and the current cost status, it is unlikely that it leads to consistent decrease in product unit cost of manufacturing or product achievement unit cost of manufacturing targets. Investigating variations should support MCI, i.e. identify deviations between manufacturing target costs and actual manufacturing costs. Even if *kaizen costing* seeks to achieve this variation, the establishment of the decreased directions of manufacturing costs is not achieved scientifically by continuous measurement of non-productivity (Matrix 1), LW (Matrix 2) behavior, and LW associated matrices (Matrix 3).

Continuous quantification of each core process of a PFC manufacturing flow in time costs (e.g. one minute cost of equipment or one minute of direct labor) and physical unit costs is performed regardless of the cost system used in the company. As can be seen in Figure 3.8, some of the manufacturing costs of a PFC main process are costs that contribute to product manufacturing (value-added costs perceived by customers), while some of the manufacturing costs are associated with the 14 categories of LW.

Table 3.6 shows an example of a monthly CLW assessment for PFC 1 in Matrix 1 with the three technological processes ("I1," "I2" and "II1"). As can be seen from the total monthly of manufacturing costs for "PFC 1" of $150,000, only $91,275 represent the cost of products manufactured in the month and the remaining $58,725 is represented by CLW. If it is taken into account that 15,000 units were made that month, then it can be stated that the average unit cost of products is $10, cost of products in manufacturing is $6,085 (about 60%), and CLW in processes and system is $3,915 (approximately 40%). To target future improvements of the CLW, Matrix 3 is prepared.

Matrix 3 is a sensitive point of the MCPD system. The process of assessing future improvement opportunities and the pressure on Matrix 1 to identify between 30% and 40% of manufacturing costs depends on the correctness of the collected level of CLW for each PFC. Therefore, based on the information collected daily in Matrix 1 and on the basis of LW monthly transformation into costs, the current level of CLW is determined. In order to increase CLW accuracy, LW conversion in costs to be achieved with a frequency of less than a month (at least for certain direct costs: raw material, components and direct labor) would be desirable.

Figure 3.9 presents Matrix 3 updated for the last previous 6–12 months.

Following the transformation of losses and waste into manufacturing costs, part of the LW received other assessments of Matrix 1, such as: some scored with one and three in Matrix 1 were scored with three and five in Matrix 3; some scored with three in Matrix 1 were scored with one in Matrix 3; and others scored with five in Matrix 1 were scored with one or three in Matrix 3. Therefore, CLW in Matrix 3 highlighted with gray had changes against the LW assessment in Matrix 1. The total change in the assessment is ten extra points (from 200 points in Matrix 1 to 210 points in Matrix 3). There is no rule against increasing the rating score in Matrix 3.

This CLW approach is in line with the principles MCPD system (see Figure 1.10):

TABLE 3.6

Example of Monthly Valuation of CLW from Total Manufacturing Costs for the Main Processes of PFC 1 (from Matrix 1)

| | Manufacturing costs for "PFC 1" | | | | | | |
| | Transformation costs | | | | Material costs | | |
	Direct labor costs	Indirect labor costs	Manufacturing overhead	Depreciation costs	Direct material costs	Indirect material costs	Total Costs
Technological processes "I1"							
Inputs Costs (previous process)	x	x	x	x	x	x	x
Inputs of CLW (previous process)	x	x	x	x	x	x	x
New Inputs of Costs in "I1"	$12.705	$2.310	$19.635	$5.775	$71.610	$3.465	$115.500
Cost of Products in Manufacturing	$7.623	$1.386	$11.781	$3.465	$42.966	$2.079	$69.300
CLW in Process & System (in "I1")	$5.082	$924	$7.854	$2.310	$28.644	$1.386	$46.200
Total costs of process "I1"	$12.705	$2.310	$19.635	$5.775	$71.610	$3.465	$115.500
Technological processes "I2"							
Inputs Costs (previous process)	$7.623	$1.386	$11.781	$3.465	$42.966	$2.079	$69.300

(Continued)

TABLE 3.6 (CONTINUED)

Example of Monthly Valuation of CLW from Total Manufacturing Costs for the Main Processes of PFC 1 (from Matrix 1)

| | Manufacturing costs for "PFC 1" | | | | | | |
| | Transformation costs | | | | Material costs | | |
	Direct labor costs	Indirect labor costs	Manufacturing overhead	Depreciation costs	Direct material costs	Indirect material costs	Total Costs
Inputs of CLW (previous process)	$5.082	$924	$7.854	$2.310	$28.644	$1.386	$46.200
New Inputs of Costs in "I2"	$2.475	$450	$3.825	$1.125	$13.950	$675	$22.500
Cost of Products in Manufacturing	$1.559	$284	$2.410	$709	$8.789	$425	$14.175
CLW in Process & System (in "I2")	$916	$167	$1.415	$416	$5.162	$250	$8.325
Total costs of process "I2"	$2.475	$450	$3.825	$1.125	$13.950	$675	$22.500
Technological processes "II1"							
Inputs Costs (previous process)	$1.559	$284	$2.410	$709	$8.789	$425	$14.175
Inputs of CLW (previous process)	$916	$167	$1.415	$416	$5.162	$250	$8.325
New Inputs of Costs in "II1"	$1.320	$240	$2.040	$600	$7.440	$360	$12.000

(Continued)

TABLE 3.6 (CONTINUED)

Example of Monthly Valuation of CLW from Total Manufacturing Costs for the Main Processes of PFC 1 (from Matrix 1)

| | Manufacturing costs for "PFC 1" | | | | | | |
| | Transformation costs | | | | Material costs | | |
	Direct labor costs	Indirect labor costs	Manufacturing overhead	Depreciation costs	Direct material costs	Indirect material costs	Total Costs
Cost of Products in Manufacturing	$858	$156	$1.326	$390	$4.836	$234	$7.800
CLW in Process & System (in "II1")	$462	$84	$714	$210	$2.604	$126	$4.200
Total costs of process "II1"	$1.320	$240	$2.040	$600	$7.440	$360	$12.000
Total Costs of Manufacturing flow for "PFC 1"							
Cost of Product ("I1"+"I2"+"II1")	$10.040	$1.826	$15.517	$4.564	$56.591	$2.738	$91.275
CLW in Processes & System ("I1"+"I2"+"II1")	$6.460	$1.175	$9.983	$2.936	$36.410	$1.762	$58.725
Total Costs of Manufacturing flow for "PFC 1"	$16.500	$3.000	$25.500	$7.500	$93.000	$4.500	$150.000

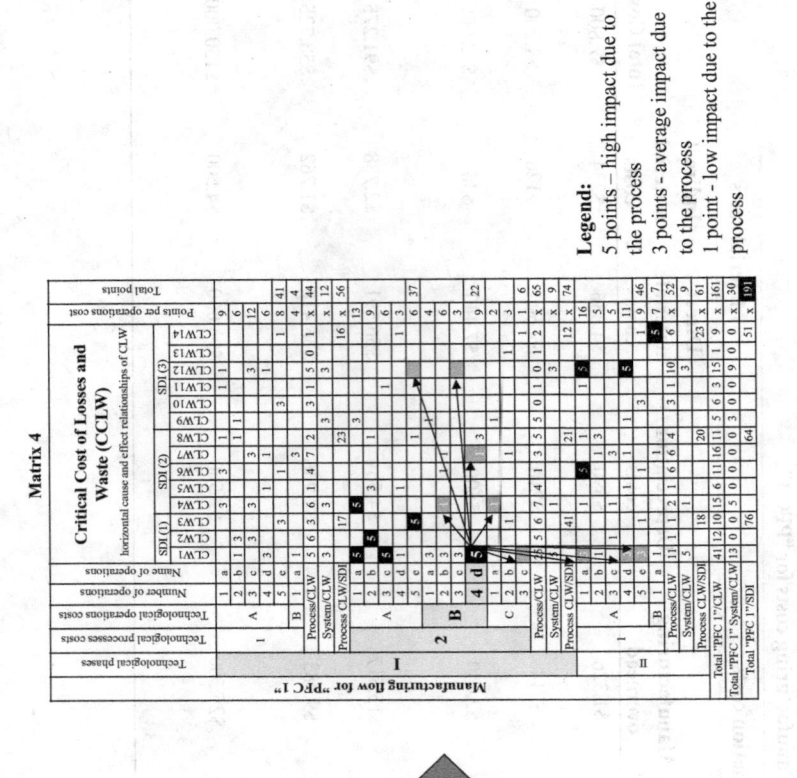

Legend:
5 points – high impact due to the process
3 points - average impact due to the process
1 point - low impact due to the process

FIGURE 3.9
Matrix 3 for measuring, analyzing and continually converting losses and waste into manufacturing costs (CLW) and Matrix 4 for analyzing and establishing critical costs of losses and waste (CCLW)–cost of losses and waste behaviors.

- Principle 3: *the continuous quantifying of losses and waste in costs for each PFC;*
- Principle 6: *coordinate all annual and multiannual systematic and systemic improvements through the need to meet annual MCI targets for each PFC).*

In order to target MCI improvements through CLW, Matrix 4 will be drawn up to establish the causal relationship between CLW. Matrix 4 is the continuation of the Matrix 2 logic from the CLW perspective. An example for the section "I2B4d" will be continued.

3.2.2.5 Matrix 4: Analyze and Establish Critical Costs of Losses and Waste

Matrix 4 establishes the acceptable future status for the total CLW of each PFC and for the overall company for the next five years and aims at identifying CCLW that sums up 80% of CLWs identified by Matrix 3 (Matrix 3 representing 30%–40% of manufacturing costs identified by LW in Matrix 1). The percentage of 80% represents the observance of the *Pareto principle* for identifying the causal relationships between losses and waste, Matrix 2, and the causal relationship between CLW, Matrix 4. Thus, as presented in Section 2.3, the acceptable future status of CLW for the following five years represents Total Offer for Cost Reduction (TOCR) (Posteucă and Sakamoto, 2017, pp. 73–75).

For example, if total costs of Manufacturing flow for "PFC 1" (see Table 3.6) is $150,000 for one month, then CLW in processes and system are $58,725 (40% of Total Costs of Manufacturing flow for "PFC 1"), and the total CCLW should be identified at the level of 80% of CLW from processes and system, i.e. $46,980 (approximately 30% of the total costs of manufacturing flow for "PFC 1").

The questions for which answers have been sought with this matrix are: *How large is the opportunity for multiannual MCI based on total CCLW? Where are multiannual MCI opportunities located based on total CCLW?* With the answers to these questions, it will be possible to establish the annual and multiannual MCI targets and priorities in addressing systematic and systemic improvements of MCI (annual and multiannual MCI means) for the next five years and, especially, for next year.

The causal relationship between CLW, which focuses on determining the CCLW, is achieved by extending the LW data in Figure 3.5, this time considering CLW. The logic of approach is similar.

Figure 3.9 shows Matrix 4. Continuing the example for "I2B4d"–LW 1, based on the data from Matrix 3 and Matrix 2, the causal relationship between CLW is identified. "I2B4d"–LW 1 becomes CCLW because it determines the occurrence of CLW at the points indicated in Matrix 4. The "I2B4d"–LW 1 effect on the Matrix 4 assessment is a decrease of the total score from 210 in Matrix 3 to 191 in Matrix 4.

Therefore, the purpose of Matrix 4 is to identify about 80% of CLW in CCLW. The ten CLW scores with five (including "I2B4d"–LW 1) are those that will attract 80% of the CLW and become CCLW. Total score in Matrix 4 becomes much lower after evaluating the ten CCLWs (as in "I2B4d"–LW 1). In this way, the ten CCLWs become the main directions for improving the manufacturing system through MCPD.

The establishing of annual and multiannual priorities for CCLW improvement is achieved with Matrix 5.

3.2.2.6 Matrix 5: Current Assumptions for Critical Costs of Losses and Waste Improvement

Once the TOCR has been set for the next five years to reduce the CLW, the annual CLW level that can be addressed by reducing the annual CCLW level for each PFC is determined (see Section 2.4). By developing current assumptions for CCLW, especially credible assumptions, it is sought to identify and consistently address the main phenomena, principles and symptoms of CLW manifestation.

Therefore, the LW of the entire manufacturing system and each of the main processes of each PFC is identified by a qualitative analysis of Matrix 1 with KPIs for LW, presented in Table 3.3 for equipment and in Table 3.4 for delivery (Posteucă and Sakamoto, 2017, pp. 116–122). Furthermore, a causality analysis was performed in Matrix 2 between LW—for example, LW in "I2B4d"–LW 1.

In Matrix 3, the example of "I2B4d"–LW 1 continued, and the conversion to manufacturing costs of equipment performance losses, namely *equipment speed down losses*, was presented. For this, we considered the cost of total minutes affected by the differences between the standard cycle time (seconds per piece) and the real-time cycle (seconds per piece) in the "I2B" process. Most of the time, equipment is a cost center. Furthermore,

CLW on *equipment speed down losses* in the cost center of the equipment is allocated to the cost objects, especially at the level of the products manufactured from that cost center.

In Matrix 4, because CLW on *equipment speed down losses* ("I2B4d"– LW 1) is a causal loss (see Figure 3.6), its effects will be seen along the entire flow. Therefore, CLW determination in the "I2B" process alone does not provide sufficient targeting and Matrix 4 is required.

Going forward, now approaching Matrix 5 for the future state of the CLW, the one minute cost of equipment speed down losses takes into account the scheduled and real-time loading of the equipment. If the overcapacity occurs for a certain period of time or *capacity planned utilization (CPU)* is lower than the *normal capacity utilization (NCU)* (see Section 1.3) then choosing a *kaizen* project to improve CLW on *equipment speed down losses* could lead to an increase in manufacturing due to the additional costs of improvement. Therefore, the increase in manufacturing capacity (SDI 1) by reducing the CLW on *equipment speed down losses* does not necessarily occur at that time due to overcapacity. The company does not need additional capacity to handle the customer order volumes at that time. Moreover, even if for "I2B4d" (LW 1–*equipment speed down losses*) the CLW from upstream processes is summed up with CLW from downstream processes and CLW on the effects of *equipment speed down losses* from the "I2B" process; more precisely, from the determination of CCLW regarding *equipment speed down losses,* it can be concluded that the CCLW approach on *equipment speed down losses* is not appropriate during overcapacity because the feasibility and opportunity of this improvement is not strategically justified at that time.

At the same time, some CLWs may remain at a sufficiently high level, perceptible especially at the shop floor level, but not included in a CCLW in Matrix 4. These types of CLW with relatively isolated behavior may become CCLW over time. Their early shop floor approach can prevent a further problem. Examples of such CLWs can be: the cost of a higher or lower consumption of lubricants in equipment; the costs of short interruptions of human labor; the cost with smaller or higher consumption of cleaning materials; the cost with smaller or higher consumption with packaging materials, etc. Such CLW types with relatively isolated behavior will contribute to CCLW creation to achieve the 80% of CLW that will need to be addressed over the next 5 years.

Under these circumstances, a CCLW strategic approach is required to meet the annual MCI targets and, implicitly, the annual MCI goal.

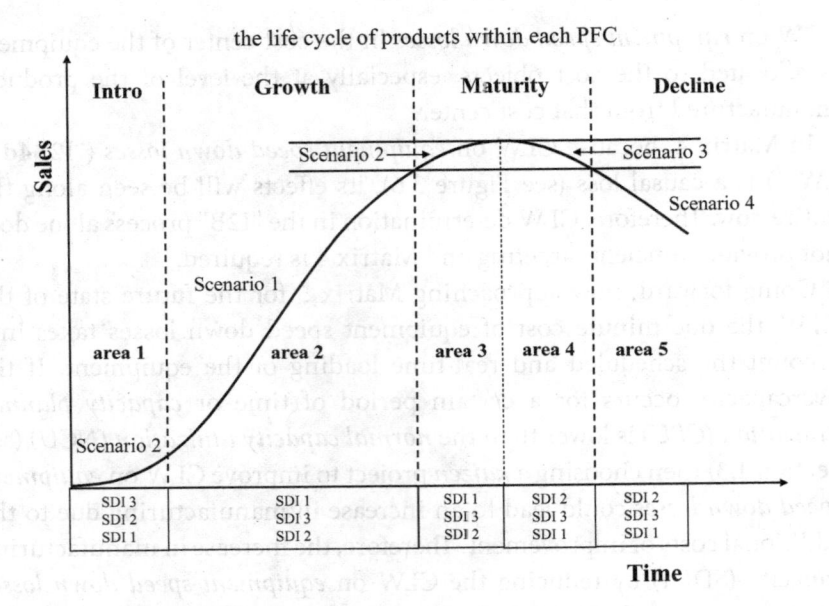

the life cycle of products within each PFC

FIGURE 3.10

Aligning directions for SDI based on product life cycle sales scenarios for a PFC.

The CCLW approach exclusively (LW causal costs and LW resulting costs along the main PFCs) will not guarantee the achievement of the annual MCI targets and a consistent decrease in product unit cost of manufacturing.

Figure 3.10 shows the alignment directions of the SDI according to the possible evolution of the sales of a product or all products of a PFC over their life cycle.

As one can see, the four scenarios described in Section 2.4 are now associated with the normal life cycle of a product or product family. It is the life cycle of products on a market with fierce competition where the "growth" area is the most consistent because products have a short life cycle.

Therefore, in each of the five life cycle areas of the products of Figure 3.10, the following SDI approaches are needed to achieve a healthy CCLW targeting continuously toward meeting the annual MCI targets:

- *Area 1* is the area in which overcapacity (CPU < NCU) and Scenario 2 (incremental growth) prevail most of the time. In this context, the order of addressing SDI is: (1) SDI 3 (Manufacturing Delivery–Minimizing Inputs–"I"); (2) SDI 2 (Manufacturing

Control–Minimizing Inputs–"I"); and (3) SDI 1 (Manufacturing Capacity–Maximizing Outputs–"E"). CCLW in the SDI 3 area will therefore be the first to be addressed in order to establish the annual and multiannual MCI targets and means, in particular by reducing delivery (CLW 9–14);

- *Area 2* is the area in which the undercapacity (CPU > NCU) or the undercapacity with a hidden overcapacity is most often encountered, amid a high level of LW that obstructs potential but unused capacity, usually expressed a low percentage of OEE and Scenario 1 (major and flexible sales growth). In this context, the order of addressing SDI is: (1) SDI 1; (2) SDI 3; and (3) SDI 2. Therefore, CCLW from area SDI 1 will be the first to be addressed in setting annual and multiannual MCI targets and means, in particular by *kaizen* projects for increasing OEE in current equipment (CLW 1) and by *kaikaku* projects to replace current equipment with new ones to meet the volume of orders (CLW 3);

- *Area 3* is the area in which the undercapacity (CPU > NCU) and Scenario 2 (incremental growth) mostly occurs. In this context, the order of addressing SDI is: (1) SDI 1; (2) SDI 3; and (3) SDI 2. Therefore, CCLW from area SDI 1 will be the first to be addressed in setting annual and multiannual MCI targets and means, in particular through *kaizen* projects for increasing OEE in current equipment (CLW 1) and by preparing for the launch of new products (CLW 2);

- *Area 4* is the area in which the overcapacity (CPU < NCU) and Scenario 3 (incremental decline) predominantly start to occur. In this context, the order of addressing SDI is: (1) SDI 2; (2) SDI 3; and (3) SDI 1. Therefore, CCLW in the SDI 2 area will be the first to be addressed in establishing the annual and multiannual MCI targets and means, especially cost of human work losses (CLW 4) and cost of raw material losses (CLW6);

- *Area 5* is the area in which the overcapacity state (CPU < NCU) and Scenario 4 (major unexpected changes–decline) predominantly start to occur. In this context, the order of addressing SDI is: (1) SDI 2; (2) SDI 3; and (3) SDI 1. Therefore, CCLW in the SDI 2 area will be the first to be addressed in establishing the annual and multiannual MCI targets and means, especially cost of human work losses (CLW 4), cost of raw material losses (CLW6) and cost of energy/utilities losses (CLW7).

Starting from this strategic alignment of CCLW with the evolution of sales, CCLW improvement directions can be established according to the context of each process of each PFC. In some cases, depending on the specifics of the companies, the shape of the product life cycle may be different. For example, for the pharmaceutical industry, *Area 1* may be merged with *Area 2* because the sales volume is very high from the very beginning. The case is similar for the cellphone industry and IT equipment industry. In other industries, the maturity area may be larger, such as in the food industry.

Moreover, because in a manufacturing company there may be more PFCs within the same manufacturing flow, strategic decisions will be made on the necessary improvements of CCLW, taking into account the need to meet the annual MCI goal by observing the seven basic principles of the MCPD system, especially of Principle No. 1: *annual and multiannual manufacturing target profit from MCI does not change at the level of the whole company* (see Table 1.10).

Following the analysis of the alignment of directions for SDI above, the current assumptions of CCLW are established and are created–Table 2.1 (see Table 2.1 and Section 2.4) is filled in:

- Credible assumptions for each MCI strategy (D1 to D8);
- Uncertain assumptions for each MCI strategy (D1 to D8);
- Vulnerable assumptions for each MCI strategy (D1 to D8).

At the same time, the cost-benefit analysis begins to identify the most feasible annual and multiannual MCI means to meet the annual and multiannual MCI targets.

Continuing the previous example, all ten CCLW identified in Matrix 4 must fit into the current SDI status for the next and subsequent years (one of the five areas in Figure 3.10). The assessment is achieved every six months to ensure that the annual MCI goal is met.

Further on, Figure 3.11 shows Matrix 5. For example, it is considered that the manufacturing company analyzed in Matrix 5 falls in the situation in *Area 2*. In this context, all five-point SDI 1 assessments are deemed appropriate to be addressed with priority, then the five-point SDI 3 and SDI 2. Continuing the example for "I2B4d"–LW 1, improving it becomes a priority for the company. Matrix 5 presents the effects of addressing solely CLW scored with five with SDI 1. In fact, each CLW in Matrix 5 is

Matrix 5

Current assumptions for CCLW improvement
(analysis of the feasibility of improvements)

Technological phases	Technological processes costs	Technological operations costs	Number of operations	Name of operations	CLW1	CLW2	CLW3	CLW4	CLW5	CLW6	CLW7	CLW8	CLW9	CLW10	CLW11	CLW12	CLW13	CLW14	Points per operations cost	Total points
					SDI (1)				SDI (2)				SDI (3)							
I — Manufacturing flow for "PFC 1"	1	A	1	a											1	1			2	
			2	b	1														1	
			3	c															0	
			4	d	1														1	
			5	e						1									2	6
		B	1	a															0	0
		Process/CLW			2	0	0	0	0	0	1	0	0	0	0	1	1	0	x	6
		System/CLW			3		1							1		1			x	6
		Process CLW/SDI					5					2						5	x	12
	2	A	1	a	**5**			1											6	
			2	b		**5**					1								6	
			3	c	**5**										1				6	
			4	d					1										2	
			5	e			**5**					1							6	26
		B	1	a						1									1	
			2	b				1		1	1								3	
			3	c															0	
			4	d	**5**							1							6	10
		C	1	a					1										1	
			2	b			1												1	
			3	c															0	2
		Process/CLW			15	5	6	3	1	1	2	2	1	0	1	0	0	1	x	38
		System/CLW			1														x	1
		Process CLW/SDI					27					9						3	x	39
II	1	A	1	a	3					3						1			7	
			2	b	1					1									2	
			3	c			1												1	
			4	d	3												5		8	
			5	e	3		1			1									5	23
		B	1	a	1							1							3	3
		Process/CLW			11	1	1	0	0	4	2	0	0	0	6			1	x	26
		System/CLW			3															3
		Process CLW/SDI					16					6						7	x	29
Total "PFC 1"/CLW					28	6	7	3	1	6	4	2	1	0	2	7	0	3	x	70
Total "PFC 1" System/CLW					7	0	0	1	0	0	0	0	1	0	0	1	0	0	x	10
Total "PFC 1"/SDI							48					17						15	x	**80**

Legend:
5 points – high impact due to the process
3 points - average impact due to the process
1 point - low impact due to the process

FIGURE 3.11
Matrix 5 for current assumptions for critical costs of losses and waste improvement.

analyzed for the area in which the company will stand in the near future (12 months).

Therefore, the answers to the following questions will be sought with Matrix 5:

- How big is it and where exactly is the opportunity for annual MCI based on annual CCLW identification located?
- What are the CLW issues to be addressed in the coming year?
- Why should those CLW issues be addressed subsequently?
- Why are the selected CCLWs a problem the next year?

3.2.2.7 Matrix 6: Setting Annual MCI Targets to Achieve Annual MCI Goal

Once the current assumptions for critical costs of losses and waste improvement for the next year (assessed every six months) have been established, annual MCI targets are set for each PFC (see Section 2.4). Matrix 6 sets annual MCI targets that converge to meet the annual MCI goal.

Matrix 6 develops together with Matrix 5 and Matrix 7, with *critical to annual manufacturing profitability tree* (CAMPT), and annual budgets for MCI (step 4 of the MCPD system).

In order to determine the order of importance of CCLW to be approached, already established in Matrix 5, the five operations of the "I2" process ("A" and "B") scored with five for the CCLW, the intensity of the annual MCI targets is set with five operations of the "I2" process depending on:

- The level of impact on the achievement of the annual MCI goal;
- The degree of difficulty of approaching improvement (the lowest are preferred); and
- The feasibility of improvement.

Therefore, Figure 3.12 presents Matrix 6 and the priorities of the five SDI 1 related operations of PFC 1.

Figure 3.13 shows an example of CAMPT structure.

When CAMPT is being developed, account is also taken of the strategic needs of productivity and quality improvements for the upcoming period, beyond the need for annual MCI goals and the need to support the annual external profitability targets (from sales).

Matrix 6

Mfg. phase	Proc. costs	Ops. costs	No. of ops	Name of ops	CLW1	CLW2	CLW3	CLW4	CLW5	CLW6	CLW7	CLW8	CLW9	CLW10	CLW11	CLW12	CLW13	CLW14	Points per operations cost	Total points
					SDI (1)			SDI (2)						SDI (3)						
I	1	A	1	a											1	1			2	
			2	b	1														1	
			3	c															0	
			4	d	1														1	
			5	e						1								1	2	6
		B	1	a															0	0
		Process/CLW			2	0	0	0	0	0	1	0	0	0	0	1	1	0	1	x	6
		System/CLW			3			1					1			1			x	6
		Process CLW/SDI					5					2						5	x	12
	2	A	1		5		5	1											6	
			2	b		5	5				1								6	
			3		5										1				6	
			4	d					1								1		2	
			5	e				5				1							6	26
		B	1	a										1					1	
			2	b				1		1	1								3	
			3	c															0	
			4	d	5							1							6	10
		C	1	a				1											1	
			2	b			1												1	
			3	c															0	2
		Process/CLW			15	5	6	3	1	1	2	2	1	0	1	0	0	1	x	38
		System/CLW			1														x	1
		Process CLW/SDI					27					9						3	x	39
II	1	A	1	a	3				3							1			7	
			2	b	1					1									2	
			3	c			1												1	
			4	d	3												5		8	
			5	e	3		1			1									5	23
		B	1	a	1							1						1	3	3
		Process/CLW			11	1	1	0	0	4	2	0	0	0	0	6	0	1	x	26
		System/CLW			3														x	3
		Process CLW/SDI					16					6						7	x	29
Total "PFC 1"/CLW					28	6	7	3	1	6	4	2	1	0	2	7	0	3	x	70
Total "PFC 1" System/CLW					7	0	0	1	0	0	0	0	1	0	0	1	0	0	x	10
Total "PFC 1"/SDI							48					17						15	x	**80**

Header note for central columns: **Annual MCI targets for all the main processes of PFC** (to fully meet the annual MCI Goal). Left side rotated label: Manufacturing flow for "PFC 1". Column labels: Technological phases; Technological processes costs; Technological operations costs; Number of operations; Name of operations; Points per operations cost; Total points.

Legend:
5 points – high impact due to the process
3 points - average impact due to the process
1 point - low impact due to the process

FIGURE 3.12
Matrix 6 for setting annual MCI targets to achieve annual MCI goal.

	1	2	3	4	5	6	7	8
	Why?	**How many? (*)**	**What?**	**Where?**	**How? (MCI means)**	**Who?**	**When?**	**How much does it cost?**
	Maximizing Outputs (SDI1)	Annual MCI target (1): xxx\$	CCLW – CLW 1	Process 3	Equipment Cycle Time Improvement	A.A.	start - end	yyy\$
	Maximizing Outputs (SDI1)	Annual MCI target (2): xxx\$	CCLW – CLW 2	Process 1	Equipment Rework Improvement	A.B.	start - end	yyy\$
	Maximizing Outputs (SDI1)	Annual MCI target (3): xxx\$	CCLW – CLW 3	Process 2	Set-up Time Improvement	A.C.	start - end	yyy\$
	Minimizing Inputs (SDI2)	Annual MCI target (4): xxx\$	CCLW – CLW 4-8	Process 1-5	Reducing Operator Handling	B.A.	start - end	yyy\$
	Minimizing Inputs (SDI3)	Annual MCI target (5): xxx\$	CCLW – CLW 9-12	Process 1-5	Energy Consumption Improvement	B.B.	start - end	yyy\$
	Minimizing Inputs (SDI3)	Annual MCI target (6): xxx\$	CCLW – CLW 13-14	Process 1-4	Implementing the Karakuri Devices	B.C.	start - end	yyy\$
Totals		(**)						(***)

Annual MCI Goal \$xxxxx

(*) as a percentage of importance in meeting the annual MCI goal

(**) total annual MCI target (xxxx\$) = annual MCI goal (xxxx\$)

(***) the total cost of annual MCI means implementations are entries in annual budgets for MCI (AMIB for existing and new products and AMCIB)

FIGURE 3.13

Example of the critical to annual manufacturing profitability tree (CAMPT).

3.2.2.8 Matrix 7: Setting Annual MCI Means to Achieve Annual MCI Targets

In order to establish the link between annual MCI means and annual MCI goals, Matrix 7 is created.

Annual MCI means are chosen to meet the annual MCI targets by fulfilling the annual basic MCI strategies of the eight change drivers of MCI described extensively in Chapter 2 (Sections 2.1.1 and 2.2.2; see Table 2.1). Examples of annual MCI means implementation, implementing the *kaizen* and *kaikaku* annual strategic projects for MCI, will be presented in Part Two and Part Three of this book.

Matrix 7 is assessed every six months and continuous planning is performed for the next twelve months. These assessments and planning are input data of AMIB and AMCIB (step 3 of the MCPD system) and action plan for MCI (step 4 of the MCPD system). The six-month assessments are to continuously achieve reconciliation between annual MCI goals and annual MCI targets and means (see Figure 3.2). As has already been said, annual MCI means targets are established with the help of Matrix 7 to set targets for LW KPIs (see Table 3.3 and Table 3.4).

Moreover, each annual MCI means (strategic *kaizen* or *kaikaku* project) is assessed at six months and 12 months after its completion in order to check the consistency over time of reaching the annual MCI targets (they are the input data for Step 6 of the MCPD system–MCI Performance).

Annual MCI means must reach the level of annual MCI targets, i.e. 6%–10% of product unit cost of manufacturing decrease (see Section 2.4). The annual impact of achieving annual MCI targets, through annual MCI means, is set at the time of the implementation of the profitability improvement project (*kaizen* and *kaikaku* for MCI) and until the end of the accounting year, then over the next period. Monitoring the actual effects of the annual MCI means is of great importance in the MCPD system. For example, if a *kaizen* improvement project to eliminate CCLW (CLW 1: equipment breakdown) lasted 11 weeks and was completed on May 4th of year N, then the benefits will be quantified by the end of year "N" (including May)–annual MCI. For each *kaizen* and *kaikaku* project for MCI, a 12-month report on its tangible and intangible effects will be prepared (see Section 2.5).

Essentially, step 6 (MCI performance) is the most important aspect of the MCPD system. Controlling the annual planning of MCI targets and MCI means to meet the annual MCI goal is essential to supporting MCPD culture at all levels of the company. The success of *kaizen* and *kaikaku* project management for MCI is continually being communicated to the company.

Therefore, all systematic (*kaizen*) and systemic (*kaikaku*) improvements to meet the annual MCI targets must begin with the need for an annual MCI goal.

Figure 3.14 shows the Matrix 7. As can be seen, a new priority of the CCLW approach has been established based on the timely planning of *kaizen* and *kaikaku* projects for MCI based on cost-benefit analysis (from CAMPT–column 8).

The final result of the seven matrices is presented in Table 3.7. It is the continuation and finalization of Table 3.5.

Therefore, by thoroughly scrutinizing this second step of the MCPD system, MCI's most sensitive challenge is addressed: setting annual targets and means. Continuous and consistent reconciliation between the top-down approach and the bottom-up approach for establishing the annual MCI goal for each PFC by developing the seven matrices is a continuous and specific activity of a distinct department of the manufacturing company. This department is responsible for planning and delivering annual and multiannual manufacturing internal profit and supporting the annual and multiannual manufacturing external profit (Posteucă, 2018, pp. 81–137 and pp. 233–244).

3.3 PHASE 2: MANUFACTURING COST POLICY DEVELOPMENT

This second phase of the MCPD system aims at coordinating MCI's means to achieve MCI targets in order to consistently support MCPD in the medium term and the long term and to ensure horizontal and vertical communication in any manufacturing company every year by developing *annual manufacturing improvement budget for existing products, a multiannual manufacturing improvement budget for new products* and an *annual manufacturing cash improvement budget (step 3: annual improvement budget development for each PFC)* and by planning, deploying and implementing the strategic *kaizen* and *kaikaku* projects for MCI *(step 4: annual action plan for MCI for each PFC)*.

Matrix 7

Manufacturing flow for "PFC 1"

Annual MCI means targets for all the main processes PFC
(to fully meet the annual MCI targets)

Technological phases	Technological processes costs	Technological operations costs	Number of operations	Name of operations	SDI (1)			SDI (2)					SDI (3)						Points per operations cost	Total points
					CLW1	CLW2	CLW3	CLW4	CLW5	CLW6	CLW7	CLW8	CLW9	CLW10	CLW11	CLW12	CLW13	CLW14		
I	1	A	1	a											1	1			2	
			2	b	1														1	
			3	c															0	
			4	d	1														1	
			5	e						1								1	2	6
		B	1	a															0	0
		Process/CLW			2	0	0	0	0	1	0	0	0	0	1	1	0	1	x	6
		System/CLW			3		1					1			1				x	6
		Process CLW/SDI					5					2						5	x	12
	2	A	1	a	5		1												6	
			2	b		5						1							6	
			3	c	5										1				6	
			4	d					1								1		2	
			5	e			5					1							6	26
		B	1	a									1						1	
			2	b					1		1	1							3	
			3	c															0	
			4	d	5							1							6	10
		C	1	a				1											1	
			2	b		1													1	
			3	c															0	2
		Process/CLW			15	5	6	3	1	1	2	2	1	0	1	0	0	1	x	38
		System/CLW			1														x	1
		Process CLW/SDI					27					9						3	x	39
II	1	A	1	a	3					3						1			7	
			2	b	1						1								2	
			3	c			1												1	
			4	d	3											5			8	
			5	e	3	1				1									5	23
		B	1	a	1						1							1	3	3
		Process/CLW			11	1	1	0	0	4	2	0	0	0	0	6	0	1	x	26
		System/CLW			3														x	3
		Process CLW/SDI					16					6						7	x	29
Total "PFC 1"/CLW					28	6	7	3	1	6	4	2	1	0	2	7	0	3	x	70
Total "PFC 1" System/CLW					7	0	0	1	0	0	0	0	1	0	0	1	0	0	x	10
Total "PFC 1"/SDI							48					17						15	x	80

FIGURE 3.14

Matrix 7 for setting annual MCI means to achieve annual MCI targets.

TABLE 3.7

Annual MCI Targets and Means (in Synthesis for a PFC)

Company Productivity Vision	What do managers want?					Years "N-1"		"N" (MCI Target)		"N" (MCI Means)			"N+1" (Target)		"N+1" (MCI Means)		
	CPM	OMIs	KPIs1	KPIs2	Unit	How much? fill in	Where? Processes/ System (*)	How much? fill in	Where? Processes/ System (*)	How much? fill in	Where? Processes/ System (*)	How?	How much? fill in	Where? Processes/ System (*)	How much? fill in	Where? Processes/ System (*)	How?
	Manufacturing Cost				$	$	P/S (***)	$	P/S	TRL/PL	P/S	K/K	$	P/S	TRL/PL	P/S	K/K
		CLW			$	$	P/S	$	P/S	TRL/PL	P/S	K/K	$	P/S	TRL/PL	P/S	K/K
			Manufacturing Capacity		$	$	P/S	$	P/S	TRL/PL	P/S	K/K	$	P/S	TRL/PL	P/S	K/K
				Cost of Equipment Losses	$	$	P/S	$	P/S	TRL/PL	P/S	K/K	$	P/S	TRL/PL	P/S	K/K
				Cost of New Equipment Losses	$	$	P/S	$	P/S	TRL/PL	P/S	K/K	$	P/S	TRL/PL	P/S	K/K
				Cost of New Product Development Losses	$	$	P/S	$	P/S	TRL/PL	P/S	K/K	$	P/S	TRL/PL	P/S	K/K
			Manufacturing Control		$	$	P/S	$	P/S	TRL/PL	P/S	K/K	$	P/S	TRL/PL	P/S	K/K
				Cost of Human Work Losses	$	$	P/S	$	P/S	TRL/PL	P/S	K/K	$	P/S	TRL/PL	P/S	K/K
				Cost of Maintenance Material Losses	$	$	P/S	$	P/S	TRL/PL	P/S	K/K	$	P/S	TRL/PL	P/S	K/K
				Cost of Raw Material Losses	$	$	P/S	$	P/S	TRL/PL	P/S	K/K	$	P/S	TRL/PL	P/S	K/K
				Cost of Energy/ Utilities Losses	$	$	P/S	$	P/S	TRL/PL	P/S	K/K	$	P/S	TRL/PL	P/S	K/K

(Continued)

TABLE 3.7 (CONTINUED)

Annual MCI Targets and Means (in Synthesis for a PFC)

Company Productivity Vision	CPM	OMIs	KPIs1	KPIs2	Unit	"N-1" How much? fill in	"N-1" Where? Processes/ System (*)	"N" (MCI Target) How much? fill in	"N" (MCI Target) Where? Processes/ System (*)	"N" (MCI Means) How much? fill in	"N" (MCI Means) Where? Processes/ System (*)	"N" (MCI Means) How?	"N+1" (Target) How much? fill in	"N+1" (Target) Where? Processes/ System (*)	"N+1" (MCI Means) How much? fill in	"N+1" (MCI Means) Where? Processes/ System (*)	"N+1" (MCI Means) How?
				Internal Logistics/ Material Handling Losses	$	$	P/S	$	P/S	TRL/PL	P/S	K/K	$	P/S	TRL/PL	P/S	K/K
			Manufacturing Delivery		$	$	P/S	$	P/S	TRL/PL	P/S	K/K	$	P/S	TRL/PL	P/S	K/K
				Cost of Waste: Raw Material Inventory	$	$	P/S	$	P/S	TRL/PL	P/S	K/K	$	P/S	TRL/PL	P/S	K/K
				Cost of Waste: Components Inventory	$	$	P/S	$	P/S	TRL/PL	P/S	K/K	$	P/S	TRL/PL	P/S	K/K
				Cost of Waste: WIP from Set-up (WIP S)	$	$	P/S	$	P/S	TRL/PL	P/S	K/K	$	P/S	TRL/PL	P/S	K/K
				Cost of Waste: WIP from Transfer (WIP T)	$	$	P/S	$	P/S	TRL/PL	P/S	K/K	$	P/S	TRL/PL	P/S	K/K
				Cost of Waste: Finished Products Inventory	$	$	P/S	$	P/S	TRL/PL	P/S	K/K	$	P/S	TRL/PL	P/S	K/K
				Cost of Waste: Packaging Inventory	$	$	P/S	$	P/S	TRL/PL	P/S	K/K	$	P/S	TRL/PL	P/S	K/K

Note (*): processes: the main processes of each PFC; system: manufacturing system approach; (**): units of measurement are determined for all PFCs by setting an average of 5–15 representative products for each PFC; (***): P/S: Process/System (manufacturing system).

Continuous targeting of MCI targets and means is achieved through this second phase of the MCPD, to achieve the long-term profit plan of both external profit (from the sale of goods and services; by increasing the number of products sold; supporting SD1, SDI2 and SDI3) and especially internal profit (from systemic and systemic improvements of MCI; continuous reduction of CCLW and CLW) through *total involvement of all departments* to achieve continuous company transformation and by reducing *reactive managerial behavior* to operational challenges based on annual and multiannual action plan for MCI (achieving annual MCI goal and CAMPT through cascading the annual action plan for MCI for each PFC).

3.3.1 Step 3: Annual Budgets for MCI

In the MCPD system logic, at the budgetary level, the conventional achievement of annual external profit requires the development of an annual master budget:

- *Operating budget* (selling budget/revenues budget, cost of goods sold budget with direct material purchases budget, with direct labor budget and with manufacturing overhead budget and selling and administrative expenses budget);
- *Financial budgets* (capital expenditure budget, cash budget);
- *Pro-forma financial statements* (budgeted income statement and budgeted balance sheet).

Going forward, in the MCPD system logic, beyond the conventional logic of external profit budgetary approach, each *operating budget* structure budget *carries CLW and CCLW (from operating manufacturing budget)*. Therefore, the MCPD system focuses on the *operating budget* (with the exception of the selling budget and administrative expenses budget), with extension over the *cash budget*. However, CLW and CCLW are also present in the case of the selling budget and administrative expenses budget, but they do not fall within the manufacturing scope. CLW and CCLW from the selling budget is the area of supply chain cost policy deployment (targets and means for cost improvement for supply and delivery).

In order to achieve the *annual MCI goal* and, implicitly, CAMPT, *annual improvement budgets* are being developed. With the help of *annual improvement budgets*, Principle No. 1 of the MCPD system is observed ("*Annual and multiannual manufacturing target profit from MCI does not change at the level of the whole company*"). Compliance with this principle (achieving the annual MCI goal), along with the annual external profit expected from sales, leads to the achievement of annual targets profit. As every structure of a manufacturing company is organized to obtain annual external profit (production; maintenance; production engineering; quality assurance; production control; cost management; development and design; sourcing and resource management; sales and research, distribution and services, etc.), the same structures need to be organized in order to achieve in parallel the internal profit from the systematic and systemic improvements of the MCI, on the basis of the continuous reduction of CCLW and CLW. Furthermore, since the CCLW approach requires a multi-process/multi-departmental approach, it is necessary to set up stable interdepartmental teams to continually have an *internal profit generation plan* converging to the annual MCI goal.

Therefore, based on the seven matrices described above, especially on the basis of Matrix 6 and Matrix 7, the *manufacturing improvement budget cycle framework* is developed in six large phases and following the *PDCA cycle* (see Figure 3.15):

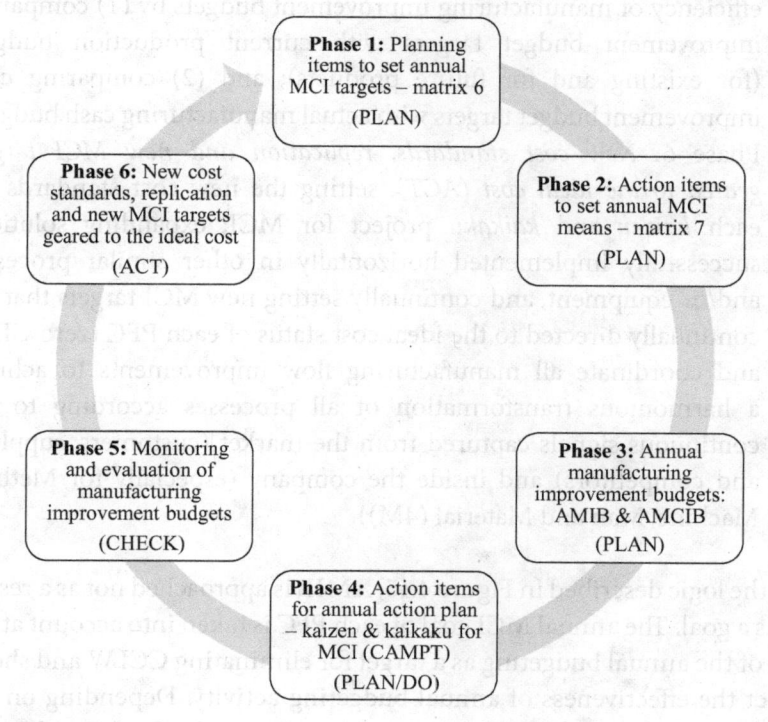

Phase 1: Planning items to set annual MCI targets – matrix 6 (PLAN)

Phase 2: Action items to set annual MCI means – matrix 7 (PLAN)

Phase 3: Annual manufacturing improvement budgets: AMIB & AMCIB (PLAN)

Phase 4: Action items for annual action plan – kaizen & kaikaku for MCI (CAMPT) (PLAN/DO)

Phase 5: Monitoring and evaluation of manufacturing improvement budgets (CHECK)

Phase 6: New cost standards, replication and new MCI targets geared to the ideal cost (ACT)

FIGURE 3.15
Manufacturing improvement budgets cycle.

- Phase 1: *Planning items to set annual MCI targets (PLAN)*: all activities from CPV and CPM to annual MCI goal and, implicitly, annual MCI targets and annual CLW and CCLW targets;
- Phase 2: *Action items to set annual MCI means (PLAN)*: activities to establish the annual MCI means and annual losses and waste targets (depending on the profitability scenarios for SDI (1); SDI (2); and SDI (3));
- Phase 3: *Annual manufacturing improvement budgets (PLAN)*: the activities to establish (1) an annual manufacturing improvement budget for existing and new products (AMIB); and (2) annual manufacturing cash improvement budget (AMCIB);
- Phase 4: *Action items for annual action plan: kaikaku* and *kaikaku for MCI (CAMPT) (PLAN/DO)*: annual actions (*kaikaku*) and activities (*kaizen*) for MCI (identifying and implementing solutions to meet MCI targets for each PFC);
- Phase 5: *Monitoring and evaluation of manufacturing improvement budgets (CHECK)*: monitoring and evaluating the effectiveness and efficiency of manufacturing improvement budgets by (1) comparing improvement budget targets with current production budgets (for existing and for future products); and (2) comparing cash improvement budget targets with actual manufacturing cash budgets;
- Phase 6: *New cost standards, replication and new MCI targets geared to the ideal cost (ACT)*: setting the new cost standards for each *kaizen* and *kaikaku* project for MCI; expanding solutions successfully implemented horizontally in other similar processes and/or equipment, and continually setting new MCI targets that are continually directed to the ideal cost status of each PFC (zero CLW) and coordinate all manufacturing flow improvements to achieve a harmonious transformation of all processes according to the continuous signals captured from the market (customers, suppliers and competitors) and inside the company (especially for Method, Machine, Man and Material (4M)).

In the logic described in Figure 4.15, AMIB is approached not as a result, but as a goal. The annual MCI goal of each PFC is taken into account at the time of the annual budgeting as a target for eliminating CCLW and shows in fact the effectiveness of annual budgeting activity. Depending on the sales scenario, external profit or internal profit may be further enhanced (Posteucă and Sakamoto, 2017, pp. 157–193; Posteucă, 2018, pp. 187–206).

3.3.2 Step 4: Action Plan for MCI

The fourth step of the MCPD system concerns the development of an annual action plan for MCI means at company level, at each PFC level and further at each department level. Annual action plan for MCI means is achieved based on CAMPT (extension of columns 5, 6, 7 and 8 in Figure 3.13). Each manager being responsible for his/her set of KPIs, based on the development of the annual action plan for MCI means, each manager will seek to achieve his/her own KPIs by planning *kaizen* and *kaikaku* for MCI. At the same time, following the logic of Figure 3.15, annual MCI action plan, the fourth phase of the manufacturing improvement budgets cycle represents the budgetary execution part of the AMIB and AMCIB, through which every person in the organization will know exactly and on a continuous basis his/her purpose in accomplishing the annual MCI goal.

Therefore, the annual MCI action plan aims at the *total involvement of all departments and beyond* and *the reduction or elimination of reactive managerial behavior,* because this step:

1. Centralizes all the annual *kaizen* and *kaikaku* strategic projects (annual MCI means)–converging to annual MCI goal;
2. Establishes the link between CCLW targets and *kaizen* and *kaikaku* targets;
3. Determines the share of *kaizen* or *kaikaku* improvement projects in CCLW reduction (e.g. 50% of CCLW);
4. Determines the share of each *kaizen* or *kaikaku* improvement project in reaching the annual MCI goal;
5. Priorities are set at one-year level within the *kaizen* and *kaikaku* annual strategic projects;
6. Establishes the team needed to be created to reach the MCI target;
7. Establishes the method and steps to address the improvement;
8. Establishes the deadline for identification of the solutions for improvement and for the implementation and validation of the solutions approved by the senior managers;
9. Establish the tasks of each person within the *kaizen* or *kaikaku* project in order to achieve the MCI target on time;
10. The amount of money (resources) that is made available to the team to meet the MCI target (at least the cost of *kaizen* or *kaikaku* team wages, but also other material costs; in fact, the cost-benefit analysis is performed in advance and then at the end of the *kaizen*

or *kaikaku* project for MCI, ensuring from the very beginning that having all resources required to meet annual MCI targets is the top management task to fulfill the internal and external profit plan) is established from the very beginning as far as possible.

The main activities of the annual MCI action plan are (Posteucă, 2018, pp. 206–211):

- Preparing an annual list of actions and activities of MCI means for each PFC and for the whole company;
- Detailed annual planning of all strategic improvement projects (all factors that may affect the development of *kaizen* and *kaikaku* projects in due time to achieve the annual MCI targets are reviewed and taken into account);
- Define, plan and prepare individual plans for all people involved in *kaizen* and *kaikaku* projects declared to be strategic to meet the annual MCI targets (each person who is part of a *kaizen* team and/ or *kaikaku* team is notified of this fact at the beginning of the year).

By continually designing and updating the annual MCI action plan, the internal strategic communication of the MCPD system, from top management to shop floor, and vice versa, based on the catchball process, is achieved. At the same time, the role of visual management is decisive in order to ensure vertical and horizontal communication.

3.4 PHASE 3: MANUFACTURING COST POLICY MANAGEMENT

The goal of manufacturing cost policy management, the third and final phase of the MCPD system, is to reduce or eliminate *incorrect and incomplete improvement project implementation* by continually targeting these improvement projects with the need to reduce manufacturing unit cost imposed by the market.

Therefore, the three steps of this last phase of the MCPD system refer to:

- Total involvement of participants within the *kaizen* and *kaikaku* projects for MCI to achieve the annual MCI goal by achieving the annual MCI targets and means, based on the annual MCI action plan developed in the previous step, based on a structured meeting of *kaizen* and *kaikaku* for MCI–MCI means *(step 5: Engage the Workforce to Achieve the MCI Targets)*;

- Continually monitor the performance level of reaching the annual MCI targets for meeting the annual MCI goal—beyond monitoring and evaluation of manufacturing improvement budgets *(step 6: MCI Performance Management)*;
- Ensure the deployment of daily MCI management at the shop floor level by: (1) monitoring KPI of MCI (see Tables 3.3–3.5) and their connections with others' KPIs (see Tables 3.1 and 3.2); (2) continuous development of management branding *(contextual managerial behavioral identity)* (Posteucă, 2011; Posteucă and Sakamoto, 2017, pp. 240–244; Posteucă, 2018, pp. 229–231); (3) continuous implementation of new standards for losses and waste and, implicitly, for costs; (4) creating replicas of viable proven solutions in a *kaizen* or *kaikaku* project to apply to other processes and/or similar equipment; and (5) continuously recreating MCI target levels so that it is targeted on a continuous basis at the ideal cost or *zero CLW* level *(step 7: Daily Manufacturing Cost Improvement Management)*.

3.4.1 Step 5: Engage the Workforce for MCI

Interdepartmental organization and lifelong learning are decisive to continuously identify and implement cost-effective improvement projects amid the backdrop of a pro-cost and pro-productivity culture such as MCPD culture.

The basic aim of this step of the MCPD system is to address CCLW through interdepartmental teams (*kaizen* and *kaikaku* teams—MCI means) to increase the objectivity and consistency of CLW improvements for each PFC, starting from the current status of CLW, fully and continuously understanding all the links between losses and waste, manufacturing costs and processes/equipment and defining and continuously becoming aware of the zero CLW status or the ideal cost of each PFC. The contextual approach of each PFC provides an increase in the chances to meet the annual MCI goal (centralized at the company level). Figure 3.16 shows the method for approaching the cardinal goal of the MCPD system, that of ideal costs or zero CLW/CCLW (implicitly of zero losses and zero waste to ensure continuous improvement of current and/or future capabilities). For example, if SDI (1)–CLW 1 (cost

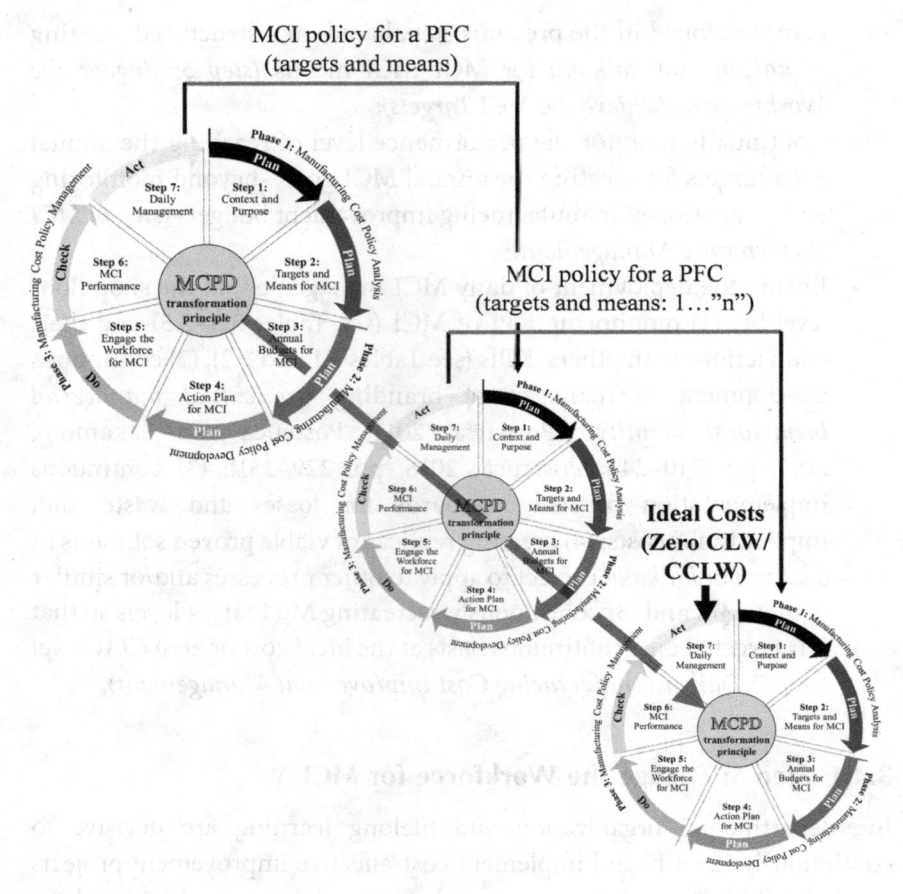

FIGURE 3.16
Successive sequences of MCPD to reach ideal costs (zero CLW/CCLW) for each PFC.

of equipment breakdown losses), cost of equipment breakdown losses has a current level of $50,000, by setting targets for successive *kaizen* projects for MCI (CLW 1–cost of equipment breakdown losses) is ideally aimed at cost or zero cost of equipment breakdown losses. The difference between the current status of CLW and the ideal state (zero CLW/CCLW) across all PFCs represents the MCPD system stake or the stakes of the harmonious transformation of manufacturing flow for uncovering hidden reserves of profitability. Therefore, each *kaizen* or *kaikaku* project for MCI is approached in a structured way, following the seven steps of the MCPD (implicitly, the PDCA cycle) on a continuous basis. Some types of *kaizen* and *kaikaku* projects for MCI will be presented in the second and third part of the book.

The basic activities of this fifth step of MCPD are:

- *Departmental and interdepartmental organization for achieving the MCI targets*: assigning tasks to each manager to meet MCI targets, besides the current operational tasks to achieve the production plan (external profit); each department manager will have a number of *kaizen* and/or *kaikaku* projects for MCI for which he/she will be responsible and for which he/she will develop departmental KPIs to continuously monitor the contextual evolution of CCLW and CLW for each PFC. This seeks to ensure that annual MCI targets will be achieved at pre-established deadlines to achieve the annual MCI goal;
- *Establishing sources for determining training needs related to the fulfillment of MCI targets*: as we have said before, the development of a culture of lifelong learning is a basic necessity in terms of supporting MCPD; from the perspective of MCPD, in line with the MCPD's basic slogan ("*Everyone, everywhere, all the time, together, in the same direction, and all the same.*"), any training and/or workshop program should address a concrete problem faced by the company or with which it may face the future or, in other words, these programs should be directly or indirectly linked to concrete elements of the CLW; identifying and prioritizing the annual training needs will focus both on supporting people's current tasks (SOP) and on MCI (improving SOP from MCI perspective);
- *Developing an initial and updated annual training plan for operators, supervisors and managers to achieve MCI targets*: each company develops an annual training plan to support MCI means targets at all levels. Examples of skills to be gained are: (1) increasing planning skills, supervision and motivation of MCPD implementation teams; (2) increasing the accuracy of setting the CLW and CCLW; (3) improving skills at setting MCI targets and means; (4) increasing delegation capacities; (5) increasing CCLW and CLW monitoring abilities; (6) improving the skill of choosing methodologies to improve processes/equipment; (7) increasing general business understanding skills; (8) increasing skills of addressing and solving a cost problem, and not only for the shop floor; (9) increasing abilities to make improvements–*kaizen*; (10) increasing the ability to collect all losses and waste, etc., accurately and within time.
- *Running the activities and actions of the annual MCI means to meet annual MCI targets*: all MCPD activities so far have been performed

to select the most profitable *kaizen* and *kaikaku* projects for MCI; identifying solutions to reduce CCLW and implementing, verifying, and expanding on the horizontal the solutions proved to be valid. This involves going through the seven steps of MCPD within each CCLW improvement project with the goal of achieving the ideal cost status or zero CLW/CCLW (see Figure 3.16).

Each *kaizen* or *kaikaku* team for MCI will receive from the MCPD steering committee the CLW or CCLW description and its implications in the company's processes and need for development, based on concrete data and information (charts, trends, positioning on layout, actual OEE, actual OLE, etc.). Afterward, the *kaizen* or *kaikaku* interdepartmental team will analyze the MCI target, set the main objectives and data needed to be collected by each member, analyze the phenomena and principles, and broadly define MCI issues and propose solutions to improve CCLW.

3.4.2 Step 6: MCI Performance Management

Annual MCI performance management refers to the effectiveness and efficiency of annual MCI means at the level of each PFC process, to check the achievement of the annual MCI targets and, implicitly, the annual MCI goal.

More specifically, annual MCI performance management refers to:

- *Performance evaluation of AMIB/AMCIB for each PFC*: AMIB/ AMCIB performance assessment refers to the degree of achievement of the annual MCI goal; annual MCI goal variation analysis is performed for each process of each PFC; the annual goal of MCI goal variation analysis is to identify new directions for reaching annual MCI targets and not just to look for the reasons why targets have not been met; from the perspective of MCPD, annual planning and re-planning of the MCI goal represents the true value of a management team within a company; AMI/AMCIB performance assessment refers to the effectiveness and efficiency of annual MCI means (effectiveness means the level of fulfillment of the annual MCI mean targets, and efficiency means the annual MCI means targets in the target level of pre-consumed consumed resources);

- *The assessment of the degree of employee involvement to achieve MCI targets*: this assessment mainly focuses on the following aspects: (1) the level of internalization of MCPD system concepts by managers; (2) continuous support of managers in reaching the annual MCI targets and means; (3) the degree of interest of new employees in their actual participation in achieving the annual MCI means targets; (4) documenting each process from the annual CLW and CCLW targets' perspective; (5) the depth and quality of lifelong learning programs and the assessment of their results obtained over time; (6) elimination of omission of negative results in annual MCI targets and means reports; (7) limiting people's participation in achieving the annual MCI means by planning too much volume of production; etc.
- *Systemic cost improvement performance management*: it evaluates the performance level beyond the AMIB and AMCIB performance assessment at the CLW and CCLW KPIs levels (see Tables 3.1–3.5) Therefore, the systemic evaluation of MCI involves understanding the phenomena and principles of manifestation of productivity and quality issues at the level of the processes of each PFC, especially *waste (stocks) elasticity on the losses.*

Therefore, MCI performance management involves pursuing the achievement of the annual MCI goal by accomplishing the annual MCI targets and means (Posteucă and Sakamoto, 2017; Posteucă, 2018).

3.4.3 Step 7: Daily MCI Management

Daily MCI management refers to the daily control of processes and equipment or, in other words, shop floor management for MCI.

This daily control implies:

- Checking the achievement level of productivity policy deployment at the level of all KPIs (see Figure 3.1 and Table 3.1 and 3.2)–to check routine tasks (to achieve the annual external profit expected from sales/production);
- Verifying the degree of achievement of MCI policy deployment for a PFC from the KPIs level related to losses and waste (see Figure 3.2 and Table 3.3, 3.4 and 3.5)–to check the tasks for improving and innovating in the workplace (to achieve annual MCI goal).

For the continuous monitoring of KPIs, the following are used: (1) MCI PFC Status Board; (2) MCI Department Status Board and (3) MCI Process Status Board (Posteucă, 2018).

The continuous promotion of the MCPD system implies the development of an atmosphere that focuses on the growth of all people's initiatives with the help of the daily MCI management process (DMCIP). DMCIP :

> "is a set of principles, processes and tools that enable the monitoring and achievement of the annual MCI targets and means at all levels of the organization and clearly establish decision-making levels by designating owners for all MCI-related KPIs [...] ensuring fast communication on the horizontal and vertical in order to have a quick reaction and an adequate response to any performance problem regarding the achievement of the annual MCI means targets on a continuous basis"

(Posteucă, 2018, p. 231).

To achieve the annual MCI goal consistently, the management message must be constant, as must the need to achieve external profit. Real engagement and managers' behaviors are extremely important. Monitoring, continuous development and improvement of managerial behavior at all levels should be in line with the expectations of employees and other people outside the company, especially in crisis situations. These desirable behaviors create the premises of a company constantly focused on learning, creativity and innovation. In the MCPD system logic, this expected managerial behavior is called *management branding (MB)* (Posteucă, 2018; Posteucă and Sakamoto, 2017). Management branding (MB)

> "is a managerial system that, by an integrated approach, creates and syn-chronizes, for the application, contextual managerial behavioral identities in order to increase organizational productivity and/or economic growth"

(Posteucă, 2011).

Therefore, the identity of managers must be pro-productive and pro-cost.

The success of the MCPD system is ensured by MCI's day-to-day management, as only strict monitoring and prompt intervention on failure to achieve the MCI targets can lead to timely responses to eliminate any variation of MCI targets.

3.5 CONCLUSIONS

Therefore, in this third chapter, emphasis was placed on the mechanism for choosing the most profitable productivity improvement projects by targeting them in line with the need for a consistent MCI. Establishing and meeting annual MCI targets by continuously and harmoniously transforming the entire manufacturing flow of each PFC and the entire company, depending on market signals, according to the need for manufacturing profit and according to the current state of the company's main processes, requires the total involvement of managers and all people within the company at all levels, and people outside the company.

Through the deployment of *kaizen* and *kaikaku* projects in a structured way and constantly directed to the ideal cost (zero CLW/CCLW and, implicitly, zero losses and zero waste to ensure a growing capacity of current and future resources), the manufacturing flow transformation is achieved through productivity. The continuous measurement of non-productivity (LW) and its transformation into costs (CLW and CCLW) provides an exact definition and localization of the manufacturing company's problems. By analyzing and continuously reducing CLW and CCLW on the basis of ongoing *kaizen* and *kaikaku* projects for MCI, month after month and year after year, both external manufacturing profit through maximizing outputs ("E") and internal manufacturing profit through minimizing inputs ("I"), regardless of the sales trend, are enhanced.

4

Annual Transformation Example through MCPD System: MCI with at Least 6% per Year for a PFC

In this chapter, a real application of the MCPD system will be presented in a manufacturing company—this is a continuation of the example from "BB-Plant" in Chapter 1: an electronics company/repeated lot, manufacturing and assembling industry, with five main manufacturing processes of a single PFC.

The main objectives of this application are to highlight the key actions and activities needed for a harmonious transformation of production flow in Scenario 1 (major and flexible sales growth; see Table 2.1 and Figure 3.10); and to provide a way to address the main challenges of consistent implementation of the MCPD system.

The company background and current business needs will be presented in Section 4.1. In Sections 4.2, 4.3 and 4.4, the implementation of the three phases and the seven steps of the MCPD system will be presented. Finally, Section 4.5 will show the tangible and intangible results, and the conclusions of the application will be given in Section 4.6.

4.1 COMPANY BACKGROUND AND BUSINESS NEEDS

Continuing with the example from "BB-Plant" in Chapter 1, the company joined the MCPD system in the "N-8" year with unsatisfactory sales volumes and annual profit, and amid increasing competition at the price level and a need to increase efficiency of investment.

"BB-Plant" uses the Productivity Business Model (PBM) to continuously support the symbiosis between productivity vision, mission and strategies and general policy deployment (OMIs, KPIs levels) and MCI policy deployment (KPIs levels, KKIs, DMIs) (Posteucă and Sakamoto, 2017, p. 21).

From the perspective of the MCPD system, the major concerns of the top management at "BB-Plant" were:

- To meet the annual and multiannual manufacturing target profit, by deeply understanding the relationship between losses, costs and costs of all the main processes of PFCs, irrespective of the trend in sales (rising or falling) and regardless of the current capacity situation (undercapacity or overcapacity); and
- To choose to scientifically and successfully implement the most profitable annual and multiannual strategic projects of systematic (*kaizen*) and systemic (*kaikaku*) improvement of profitability by increasing productivity.

Top management at "BB-Plant" was aware that any business strategy, no matter how brilliant it might be, without an irreproachable execution could become tortuous and with many losses and wasted opportunities more or less perceptible and quantifiable.

At the same time, top management at "BB-Plant" was aware that the current performance could be improved by:

- Increasing the volume of outflows from current processes both for the manufacturing of products and for delivering them;
- Increasing quality level; and above all,
- Reducing the manufacturing unit costs to meet the real customer demands.

Even if "BB-Plant" has joined the MCPD for some time, the MCI opportunities exist amid the need to adapt manufacturing flow to market needs.

Note: all the following money figures are in *thousands of dollars*.

Major demands for the "N-1" year (strategic goals to support the mission and the vision of productivity), stemming from the departments of marketing, sales, production planning, maintenance and production, were:

1. Achieving the annual manufacturing target profit of $37,000 by achieving the annual manufacturing external target profit (from sales) of $32,000 and an annual manufacturing internal target profit of $5,000 or annual MCI goal;
2. Price decrease by reducing manufacturing unit cost by $1.5 per unit;
3. Increasing sales volumes to $250,000 (from $190,000 in "N-2," including the situation of decrease of manufacturing unit cost of $1.5 per unit see Figure 1.3);
4. Ensuring the achievement of a production number (PN) of 3,265,000 units;
5. Ensuring a capacity planned utilization (CPU) of 3,275,000 units (including by increasing effectiveness by developing improvement projects in "N-1" year to increase OEE equipment considered to be strategic, see Figure 1.4);
6. Minimum achievement of MCI target of 6% or annual manufacturing internal target profit rounded to $5,000 of the total cost of manufacturing of $72,975 (rounded to $73,000). This amount of $5,000 was projected at year-end "N-2" and did not change significantly over the year "N-1." It was considered that the annual stake of 6% MCI target will have positive effects in the coming years along with the achievement of the MCI target in the coming years;
7. Identifying and addressing CLW and CCLW targets and means at the level of the five main PFC processes to achieve the 6% ($5,000) MCI target through systematic (*kaizen*) and systemic (*kaikaku*) improvements;
8. Launching two new products in year "N-1" to support sales volumes at $250,000 by 3.7%.

The method of applying the three phases and seven steps of the MCPD system is presented hereinafter to meet the goals required above by tying all the *kaizen* and *kaikaku* improvements to the need to achieve a minimum MCI target of 6%.

4.2 PHASE 1: MANUFACTURING COST POLICY ANALYSIS

The purpose of this first phase of the MCPD system at "BB-Plant" was to define the issue of the annual MCI goal and to identify the root causes of CLW throughout the entire manufacturing flow (annual CCLW to support the annual MCI targets and furthermore the annual MCI goal) to achieve annual and multiannual manufacturing target profits by translating the manufacturing unit cost strategy imposed by market competition into action (alignment for setting the annual MCI targets and means or management by MCI targets and means/management by cost policy).

4.2.1 Step 1: Context and Purpose of MCI

Since joining the MCPD system in the year "N-8," "BB-Plant" has developed the symbiosis between the general policy deployment (see Figure 3.1) and MCI policy deployment (see Figure 3.2) to support its vision and mission through MCPD system. General policy deployment addresses productivity vision translation at KPIs level for the main processes of PFC. The five annual and multiannual productivity vision drivers that aim to maximize outputs (maximize efficiency; reducing losses especially by increasing OEE equipment) and minimize inputs (maximizing efficiency; in reality, by minimizing waste) (Posteucă, 2018, p. 142), are:

- Manufacturing profit increase;
- Safety, health and environmental increase;
- Product number increase (*critical driver to achieve manufacturing profit through effectiveness*; (1) by increasing capacity by increasing OEE and (2) by reducing delivery times, more specifically, through total factory lead time synchronized to takt time);
- Manufacturing unit cost decrease (*critical driver to achieve manufacturing profit through efficiency*); and
- Morale increase (to increase the real involvement of all people within and outside the "BB-Plant" in identifying, choosing, implementing and monitoring profitable productivity improvement projects coordinated by MCI targets and means).

The five annual and multiannual drivers of productivity have been detailed at the level of manufacturing (Posteucă, 2018, pp. 85–127). Table 1.1 presents the context and purpose of the need for MCI to be applied to "BB-Plant" in the predicted way for the year "N-1." As one can see, the MCPD system directly captures the CLW level and the annual MCI ($5,000) or annual MCI goals stakes in the profit and loss statement.

The MCPD analysis has gone from the current stage of CLW and CCLW (the annual and multiannual offer for MCI) and from the current and future capacity analysis required to support the annual manufacturing target profit for the year "N-1" of $37,000.

As can be seen in Table 4.1, just like every year, the MCPD profit and loss statement for year "N-1" was the annual target number of units to be sold and the average annual target selling price. Both the annual target number of units to be sold (sought to be as large and as accurate as possible and taking into account the normal capacity utilization (NCU)) and the

TABLE 4.1

Example of the MCPD Profit and Loss Statement (from January 1 to December 31 of "N-1" for "BB-Plant")

MCPD pro-forma (forecast) profit and loss statement (unit: $000)	
Sales	**$250.000**
Annual Target Selling Price (average)	$76,6
Annual Target Number of Units to be Sold	$3.264
Cost of Manufacturing	**$73.000**
Cost of Losses and Waste (CLW) Identified in Manufacturing	**$28.500**
Actual Cost of Losses	**$20.000**
Cost of Equipment Losses	$12.000
Cost of Human Work Losses	$5.000
Cost of Material Losses	$2.000
Cost of Energy/Utilities Losses	$1.000
Actual Cost of Waste	**$8.500**
Cost of Waste: WIP Set-up	$850
Cost of Waste: WIP Transfer	$2.975
Cost of Waste: Finished Products Stock	$425
Cost of Waste: Raw Materials Stock	$2.805
Cost of Waste: Components Stock	$850
Cost of Waste: Packaging Stock	$595
Cost of Finished Product (with added value)	**$44.500**
Cost of Losses and Waste (CLW) Feasible to Improve	**$26.500**
Annual Critical Target Cost of Losses and Waste (CCLW)	**$5.000**
Annual Critical Target Cost of Losses	$3.150
Annual Critical Target Cost of Waste	$1.850
Gross Profit in Manufacturing	**$177.000**
Selling, General and Administrative Expenses	**$145.000**
Annual Manufacturing External Target Profit (from sales)	**$32.000**
Annual Manufacturing Internal Target Profit	**$5.000**
Annual Manufacturing Target Profit	**$37.000**

required average level of the annual target selling price (sought to be as small and as acceptable as possible by the target customer typology aimed for each market and for each geographical area, taking into account the internal multiannual strategy for manufacturing unit cost improvement) are set by the marketing department.

Furthermore, from the total manufacturing costs of $73,000 forecast for "N-1," the CLW level identified at the level of five manufacturing process of PFC is established at $28,500, consisting of the actual cost of losses of $20,000 and actual cost of waste of $8,500. The amount of $28,500 represents the current total offer for annual and multiannual MCI, and the amount of $44,500 represents the cost of the finished product with value added (costs considered acceptable to be paid by customers).

Furthermore, the total CLW considered feasible to be improved at that time (at the end of the year "N-2" for year "N-1") was of $26,500 (93% of the total CLW of $28,500).

In order to achieve an annual production target of $37,000, especially the $5,000 MCI annual stake, the MCPD team together with the top management at "BB-Plant" initially chose nine strategic projects to improve the annual productivity systematically (*kaizen*) and one strategic improvement project (*kaikaku*) for new equipment to support the timely launch of the two new products planned to be launched in the year "N-1."

At the same time, the exact contribution of each productivity improvement project to the annual manufacturing internal target profit of $5,000 and an annual critical target costs of losses of $3,150, respectively, was determined, following the need for preponderant increase of outputs/ capacity (effectiveness; losses improvement and OEE increase for strategic equipment) and annual critical target costs of waste of $1,850 by addressing convenient opportunities to reduce inputs that were easy to achieve and had high impact, and by improving efficiency (waste and losses, especially all types of inventory).

During the first half of the year "N-1" there were a series of events that led the MCPD team and the top management of "BB-Plant" to decide between:

1. The initiation of two *kaizen* projects to increase OEE to reach the annual manufacturing target profit of $37,000 in particular by reducing the cost of consumables from suppliers; and
2. The pressure increase in annual MCI means targets of the 9 *kaizen* strategic projects.

The final decision was to increase the pressure on annual MCI means targets. This decision was necessary in order not to violate Principle No. 1 of the MCPD (not achieving the annual manufacturing target profit of $37,000).

With the help of the MCPD system, for the reduction or elimination of the degree of uncertainty on the achievement of the annual manufacturing profit of $37,000, the most certain systematic (*kaizen*) improvement projects were identified, as well as the systemic (*kaikaku*) improvement project to contribute to achieving MCI targets in a feasible way, as every year. The total involvement of all people in the company and the MCPD consultant has been decisive in the context of a consistent and accurately-measured amount of CLW and CCLW in the main PFC processes.

To comply with this essential principle of MCPD, i.e. to achieve the annual manufacturing profit, all items in Table 1.1 have been subject to successive adjustments by the MCPD team and top management during the year "N-1," especially in the context of a growth forecast of sales for the year "N" (increase of capacities and distribution forces).

For the year "N-1," most of the sales had already been contracted at the beginning of the year "N-1" and there were no significant differences in activity volumes.

For year "N," starting from the current state of year "N-1," as sales tended to continue their upward trend, as predicted in the budget of the year "N," the PFC effectiveness scenarios were drafted more specifically for the increase of OEE and outputs (increase of current equipment capacities), as well as scenarios for increasing efficiency, namely reducing overall manufacturing costs (especially for payroll costs and material costs) until the annual manufacturing profit of $40,000 can be achieved both by annual manufacturing external target profit (from sales) and especially by annual manufacturing internal target profit (annual MCI goal).

4.2.2 Step 2: Annual MCI Targets and Means

In order to present the CLW level and the annual stake for MCI ($5,000; annual MCI goals) directly in the profit and loss statement, continuous measurements of losses and waste and associated CLWs for the five main PFC processes are carried out to determine thereafter the MCI targets and means to meet the annual MCI goal of at least 6%. The purpose of this step is to set improvement priorities and to choose what will be done in the year "N-1" to achieve the MCI target of 6% and what will remain to be performed in the future. For this, the following activities were carried out, concentrated in the seven matrices described more broadly in the previous chapter:

- *Matrix 1*: continuous measurement of losses and waste KPIs for all main processes of PFC. The main purpose of this matrix is to know the current state of losses and waste in the year "N-1," based on "N-2" data and future projections. The measurements were performed at the level of the five main PFC processes;
- *Matrix 2*: analyze and identify sources that generate losses and waste across the major processes for PFC. The main purpose of this matrix was to identify the root causes of losses and waste along the manufacturing flow for PFC at "BB-Plant" in year "N-2" and which had an impact on the predicted state in year "N-1." Matrix 2 was generated with the help of *losses and waste stratification flow analysis (LWSFA)* and 15 main sources of losses and waste generation were identified at the process level upstream and downstream of the entire manufacturing flow, of which nine were chosen as the most important and easily addressable, called critical losses and waste (e.g. the impact on losses and waste of breakdown/breakdown losses of equipment in process 3 across the whole PFC). There was a desire to abide by the Pareto principle in a transversal manner throughout the entire manufacturing flow of PFC (roughly 80% of the effects come from 20% of the causes);
- *Matrix 3*: measuring, analyzing and continually converting losses and waste into manufacturing costs. The primary purpose of this matrix was to convert to CLW all the losses and waste discovered at the level of the five main PFC processes with a total value of $28,500 (the current total offer for the annual and multiannual MCI identified in year "N-2" and with projections in year "N-1"; usually CLW ranges between 30%–40% of total cost of manufacturing).

Fixed, and above all, variable costs were allocated to five cost centers, the five main processes of the PFC, using three allocation methods ("BB-Plant" used normal costing in year "N-1"—actual cost of materials, actual cost of labor and a predetermined overhead rate; it subsequently switched to standard costing and setting targets for annual costs to improve standard costing):

- *Allocation on the basis of real data*: the allocation basis was man-hours by product for direct labor costs (regular and part-timer employee; variable cost), and the allocation basis for direct material costs (variable cost) was the actual consumption per unit;

- *Allocation on the basis of monthly data* (as accurate as possible): energy (consumption-based allocation; electricity and fuel; variable cost, depreciation (operating time-based allocation; straight-line method; fixed costs);
- *Apportionment on the basis of monthly data*: the allocation basis was predetermined for maintenance/spare parts costs (fixed/ variable costs), tool costs (variable costs), die and jig costs (variable costs), indirect labor costs (fixed/variable costs) and indirect material costs/auxiliary material costs (fixed/variable costs).

Following this allocation of the fixed costs and, especially, the variables on the five cost centers, the five main processes of the PFC, which aimed at a more accurate allocation of the overheads at each process level (e.g. energy and depreciation; the effort to locate as accurately as possible the overhead at each process level continued in the years to come), the cost of the PFC-related products was determined at the amount of $73,000.

The cost structure of products was made up of:

1. Transformation cost ($25,550; 35% from cost of manufacturing of $73,000): direct labor cost, indirect labor cost, manufacturing overheads—maintenance costs/spare parts costs, utility costs, tool costs, die and jig costs and depreciation costs; and
2. Material costs ($47,450; 65% from cost of manufacturing of $73,000): direct material costs and indirect material costs.

For each of the losses and waste identified in the five PFC processes in year "N-2" and with projections for the year "N-1," potential CLWs were determined (based on the loss, waste and cost trends at process level and on the basis of planned volumes). The CLW structure for all five PFC processes, based on Matrix 1 data, was of $20,000 for current cost of losses and $8,500 for actual cost of waste (see Table 1.1). For capacity (*time-related loss–TRL*), all minutes of each type of loss identified and/or with a potential of occurrence in the process and/or system have been converted into costs, and for delivery (*physical loss and waste–PLW*) all minutes and material units associated with losses and waste identified and/or with a potential of occurrence in the process and/or system have been converted into costs.

- *Matrix 4*: analyze and establish critical costs of losses and waste (CCLW) for PFC. The purpose of this matrix is to determine the CCLW ($5,000; annual MCI goal) related to sources that generate the losses and waste across major processes for PFCs (determined in Matrix 2; the nine main sources for generation of losses and waste upstream and downstream identified through the entire manufacturing flow and called critical losses and waste), by analyzing the feasibility of each opportunity to approach CLW (93% of the total CLW of $28,500 and $26,500) and, implicitly, the feasibility of each CCLW. With the help of this matrix, the development of *critical to annual manufacturing profitability tree (CAMPT)* is started in order to accurately locate all opportunities for the annual CCLW at the level of the five main processes of PFC;
- *Matrix 5*: Scenarios for CCLW improvement for the main processes of PFC. The purpose of this matrix was to start establishing the future state of the CLW by carrying out successive simulations of transformation of process and system critical points, to help set targets for MCI and simulations on the succession of the order of approach to future *kaizen* projects and the *kaikaku* project for MCI (annual MCI means). Three target versions for each type of CCLW were developed along the five main PFC processes. CCLW aimed at the aggregated effects of the root cause of CLW triggering CLW along the manufacturing flow for PFC (in the upstream processes of the CLW generating process, in the CLW generating process and downstream of the CLW generating process). The basic logic for developing scenarios for a CCLW was: if the CCLW in the "X" process (the CLW generating process) would decrease by "Y%" (implicitly the decrease of losses of waste along the five main processes), then the impact on achieving the annual MCI target will be "$Z." The simulation of the scenarios did not stop until an MCI target of at least 6% ($5,000) had been reached under the conditions of the best possible feasibility at that time of the future *kaizen* and *kaikaku* projects (the improvement projects were selected without significant additional costs—including the equipment replacement project (*kaikaku*), by choosing equipment that is cheap, but good enough to meet current and future capacity needs

of the company's production flow). Following the continuation of simulations throughout the entire year "N-1," two adjustments to MCI means initial targets were required to meet MCPD Principle No. 1 (achieving the annual manufacturing target profit of $37,000). Matrix 5 is performed concomitantly with Matrices 4, 6 and 7 and MCPD Phase 2 and it targets CAMPT;

- *Matrix 6*: Annual MCI Targets for all the main processes of all product families. With the help of this Matrix, annual MCI targets were set for each of the five main PFC processes. Establishing annual MCI targets is the most important point of the MCPD system because MCI targets must be accepted and achieved with considerable effort by all people in "BB-Plant" and beyond "BB-Plant";
- *Matrix 7*: Annual MCI means targets for all the main processes of all product families. With the help of this Matrix, MCI means have been established to achieve accurate and timely MCI targets for each major PFC process. Annual MCI means targeted the selection of the nine *kaizen* projects and the *kaikaku* project, setting targets for process losses and waste in order to achieve an accurate and timely annual MCI goal.

The work to create Matrices 1–7 is considered completed when the CAMPT is completed (internal order of manufacturing profit by increasing productivity) and the exact sources of achieving the annual MCI target of at least 6% have been identified at the level of processes and activities, together with MCPD Phase 2 activities (MCI targets and means feasibility with the help of annual budgets for MCI and an annual action plan for MCI means).

Figure 4.1 shows CAMPT at "BB-Plant" for year "N-1."

"BB-Plant" usually sets MCI targets and means for at least 25% of the total CLW for the current year and for the next three years—the productivity master plan (Posteucă and Sakamoto, 2017, p. 21). In our example, it is about the annual and multiannual plan of manufacturing internal profit, from all five main PFC processes, to substantially reduce the amount of $28,500, which represents 39% of the total cost of manufacturing of $73,000). A minimum of 25% is considered achievable by planning a minimum of 6% of the annual MCI gol for the next four years (annual "N," "N + 1," "N + 2" and "N + 3").

This process for establishing the annual and multiannual for the manufacturing internal target profit is correlated with the annual and

	1	2	3	4	5	6	7	8
	Why?	How many?	What?	Where?	How? (MCI means)	Who?	When?	How much does it cost?
Annual MCI Goal 5000$	Maximizing Outputs	MCI target: 662$	Critical Cost of Equipment Losses	Process 3	Equipment Cycle Time Improvement	A.A.	15.01."N-1"- 15.05."N-1"	2.750$
	Maximizing Outputs	MCI target: 315$	Critical Cost of Equipment Losses	Process 1	Equipment Rework Improvement	A.B.	01.02."N-1"- 15.06."N-1"	1.560$
	Maximizing Outputs	MCI target: 378$	Critical Cost of Equipment Losses	Process 2	Set-up Time Improvement	A.C.	15.02."N-1"- 15.05."N-1"	5.650$
	Maximizing Outputs	MCI target: 725$	Critical Cost of Equipment Losses	Process 2	Equipment Breakdown Improvement	A.D.	15.01."N-1"- 15.08."N-1"	3.120$
	Maximizing Outputs	MCI target: 347$	Critical Cost of Equipment Losses	Process 3	Equipment Shutdown Improvement	A.E.	01.03."N-1"- 15.09."N-1"	1.850$
	Maximizing Outputs	MCI target: 252$	Critical Cost of Equipment Losses	Process 4	Equipment Scrap Improvement	A.F.	10.05."N-1"- 10.01."N"	2.300$
	Maximizing Outputs	MCI target: 470$	Critical Cost of Equipment losses	Process 5	Equipment Replacement	A.G.	20.01."N-1"- 30.06."N-1"	85.920$
	Minimizing Inputs	MCI target: 1388$	Critical Cost of Labour Losses	Process 1-5	Reducing Operator Handling	B.A.	20.04."N-1"- 20.02."N"	1.550$
	Minimizing Inputs	MCI target: 278$	Critical Cost of Utilities Losses	Process 1-5	Energy Consumption Improvement	B.B.	10.03."N-1"- 30.10."N-1"	12.500$
	Minimizing Inputs	MCI target: 185$	Critical Cost of Waste (WIP Trasfer)	Process 1-4	Implementing the Karakuri Devices	B.C.	20.03."N-1"- 15.11."N-1"	7.800$

FIGURE 4.1

Critical to Annual Manufacturing Profitability Tree (CAMPT) at "BB-Plant."

multiannual manufacturing external target profit by establishing the necessary capacity to be ensured through:

- Deblocking the current state (*kaizen* for MCI);
- Procurement of new capacities (*kaikaku* for MCI).

At the same time, it is intended to achieve:

- The required quality level;
- The required available time of equipment to manufacture a production number;
- The level of synchronization between the internal processes (factory lead time and, in particular, manufacturing lead time) and the market needs (takt time); and
- The required level of manufacturing unit cost (average for PFC).

In fact, on an annual basis, a reconciliation is achieved between the external target profit KPIs and the internal target profit KPIs at the time of setting CAMPT (the moment of establishing the annual MCI policy deployment of the annual MCI targets and means).

4.3 PHASE 2: MANUFACTURING COST POLICY DEVELOPMENT

The purpose of the second phase was to coordinate the annual MCI means through the need to achieve the annual MCI targets; in fact, fulfilling the CAMPT.

4.3.1 Step 3: Annual Budgets for MCI

Continuing the example from "BB-Plant" from the perspective of the MCPD system, the budget cost efficiency for the year "N-1" represents the exact achievement of the annual MCI goal of $5,000 MCI. Depending on the evolution of sales volumes, for the year "N-1," they are sharply growing (from $190,000 in year "N-2" to $250,000 in year "N-1"), the participation share of annual manufacturing target profit is established based on scenarios developed in Matrix 5 for the achievement of the annual MCI goal of $5,000 for:

- *Annual manufacturing external target profit (from sales)*: effectiveness improvement (increase of outputs/a capacity planned utilization (CPU); increase of gross profit in manufacturing on account of sales; decrease of losses, especially for equipment by increasing for the "N-1" year of the Overall Equipment Effectiveness (OEE) for the strategic equipment of the five processes and, implicitly, the Overall Line Effectiveness (OLE) from 37% to 41%), and
- *Annual manufacturing internal target profit*: efficiency improvement (decrease of inputs; increase of gross profit in manufacturing especially due to improving the variable costs, both direct and especially indirect at the level of each center of cost/processes; in order to achieve the unit manufacturing costs improvement, it is necessary to improve transformation costs and material costs);

For "BB-Plant," for the year "N-1," the percentages of contribution to achieving the annual MCI goal of $5,000 (annual MCI target of 6%) were:

- From the annual manufacturing external target profit (from sales) of 63% (namely $3,150–SDI1); and
- From the annual manufacturing internal target profit of 37% (namely $1,850–SD3 and SDI2).

Figure 4.2 shows the dynamic evolution of the method of establishing the participation scenarios to the annual manufacturing target profit of the annual manufacturing external target profit (from sales) and the annual manufacturing internal target profit for "BB-Plant" based on annual and multiannual forecasts of the evolution of sales (see Figure 1.3) and capacity planned utilization (CPU) (see Figure 1.4).

These annual scenarios of the external and internal manufacturing profit, linked with annual evolution of sales and CPU, were detailed for the five basic PFC processes at each CLW and annual CCLW in Matrix 5 to determine thereafter the exact annual MCI targets and means in Matrix 6 and Matrix 7.

Over time, even though the CLW feasible access and feasibility opportunities are getting smaller, due to the advancement of the MCPD program, the MCPD team and "BB-Plant" top management have tried to identify scientifically the annual percentage of 6% for MCI (by supporting external profit—including through the continuous launch

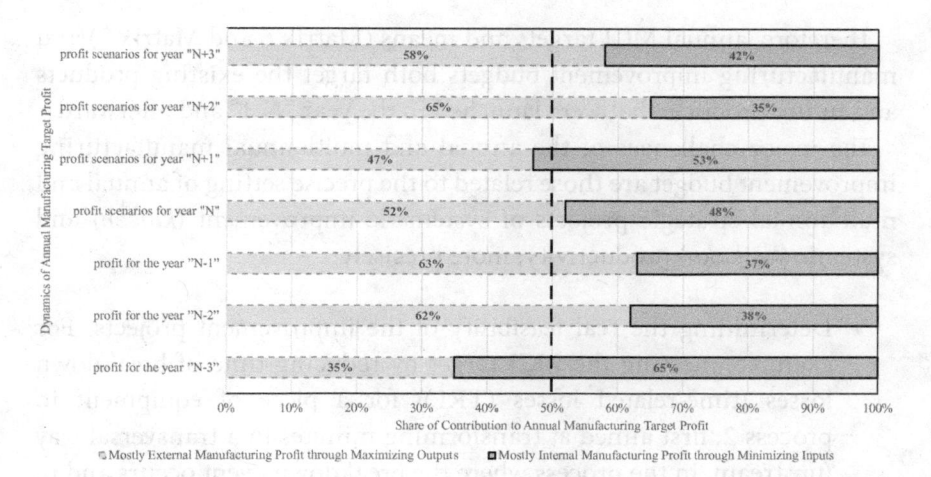

FIGURE 4.2
Establishing annual and multiannual manufacturing target profit scenarios based on the participation share of external and internal manufacturing profit

of new products—and through internal profit—continuous decrease of transformation and material costs).

Starting from Matrices 1, 2 and 3 concomitantly with Matrices 4, 5, 6 and 7, "BB-Plant" has developed three types of manufacturing improvement budgets for "N-1" respectively:

- Annual manufacturing improvement budget for existing products;
- Multiannual manufacturing improvement budget for new products (for the two new products launched in year "N-1" and for the other new products in different stages of development);
- Annual manufacturing cash improvement budget.

By identifying CLW and CCLW (Matrices 1, 2, 3 and 4) and based on current assumptions for CCLW improvement (Matrix 5), "BB-Plant" was able to identify fixed and variable manufacturing CLW and CCLW at each cost center, as a component part of the conventional budget and transformation and material costs of losses and waste from the level of each PFC product (Posteucă, 2018, pp. 35–37). In this way, the need for manufacturing unit cost improvement of $1.5 per unit was linked to the annual MCI targets and means (Matrix 6 and Matrix 7) to achieve the annual $5,000 MCI goal.

Therefore, annual MCI targets and means (Matrix 6 and Matrix 7) and manufacturing improvement budgets both target the existing products and future products that were launched in the year "N-1" and afterward.

The major challenges of the annual and multiannual manufacturing improvement budget are those related to the precise setting of annual and multiannual strategic projects of systematic improvement (*kaizen*) and systemic (*kaikaku*) productivity; more precisely:

- Determining the real feasibility of the improvement projects. For example, meeting the MCI target by reducing time of breakdown losses (time-related losses (TRL)) for a piece of equipment in process 2, first aimed at transforming minutes in a transversal way (upstream, in the process where the breakdown event occurs and in downstream thereof) with the average monthly time of breakdown losses in costs (all variable costs and part of fixed costs—especially equipment depreciation costs and part of direct labor costs) and then reducing this time with breakdown losses by achieving MCI means target (*kaizen* for breakdown) was considered to cause the achievement of a part of the annual MCI goal of $5,000 by reaching the annual target MCI targets of $725 CCLW. However, this way of thinking had some drawbacks within "BB-Plant" because the need for manufacturing unit cost improvement of $1.5 per unit was not guaranteed. When the overcapacity state (PCU<NCU) occurs for a particular process over a certain period of time, manufacturing unit cost may increase due to unnecessary or premature improvement in breakdown losses and, implicitly, the cost of breakdown losses. At the same time, the allocation of resources for such unnecessary improvement for that moment will generate a lost opportunity cost due to the failure to perform the productive current activities by the participants in the *kaizen* project for breakdown (manufacturing external profit from the production not achieved at that time for the improvement). Furthermore, even if the *kaizen* project for breakdown can be declared strategic and planned, it can still generate a series of unplanned waste, such as: WIP for transfer over the standard level, raw material stocks/costs above the standard level, components stocks/costs above the standard level, spare stock stocks above the standard level and packaging stocks above the standard level. However, if the critical cost of breakdown losses summed up along the entire PFC is significant—in our case $725 for "N-1"—then

it can be considered a feasible strategic *kaizen* project to achieve the MCI target even if there is no need for an increase in capacity in the process in which the breakdown event occurred. The level of CCLW in Matrix 4 prevails in the selection of annual MCI targets;

- Determine the period of recovery of the total resources related to the improvements (material and wage expenditures for all the time spent by people in the implementation of the improvement procedures: the last column in Figure 4.1); and
- Opportunity cost (all resources involved in improvements versus current and future gains from improvements compared to allocating those resources to improvements for the current activity to achieve production; in fact, comparing the annual and multiannual profitability of allocating the same resources for domestic profit versus allocating the same resources for current external profit).

The three types of manufacturing improvement budgets for "N-1" were assessed by comparing the level of achievement of improvement budget targets for each of the five PFC processes with the actual manufacturing budget and by comparing cash improvement budget targets with the actual manufacturing cash budget in year "N-1" and in subsequent years in Step 6 of the MCPD system (MCI Performance Management).

4.3.2 Step 4: Action Plan for MCI

After the completion of Step 2 and Step 3 of MCPD, "BB-Plant" developed the following:

- List of actions and activities for MCI means for each of the five processes;
- Planning of MCI means (defining in detail the improvement projects selected in Matrix 5, establishing the necessary results with deadlines, setting the resources assigned with deadlines, how to periodically report the status of the *kaizen* and *kaikaku* projects and how to evaluate the progress of the annual *action plan for MCI*— detailing CAMPT for rigorous implementation);
- Definition and preparation of individual plans for MCI.

4.4 PHASE 3: MANUFACTURING COST POLICY MANAGEMENT

The purpose of this third and last phase of MCPD is to achieve an effective and efficient management of the annual MCI means to meet the annual MCI targets ($5,000), consistent with the core slogan of the MCPD system: Everyone, everywhere, all the time, together, in the same direction and the same.

4.4.1 Step 5: Engage the Workforce for MCI

By this step "BB-Plant" has ensured the continuous involvement of all people in achieving the annual MCI goal of $5,000.

Specifically, the following have been achieved:

- Interdepartmental organization to address *kaizen* and *kaikaku* projects;
- Sources for determining training needs related to the fulfillment of the annual MCI targets: training needs have been identified for systematic individual improvements (*kaizen*—training examples: breakdown analysis, cycle time reduction, set-up time reduction, scrap and rework reduction, line balancing, time and motion study, layout improvement, material saving method, etc.) and for systemic improvements (*kaikaku*—for example: methods design concept (MDC), operator-level maintenance, preventative and predictive maintenance, standard costing and scientific setting of cost reduction targets (analysis of monthly variations between current cost status and target costs), etc.;
- Initial and updated annual training plan to achieve the annual MCI targets;
- Continuous monitoring of *kaizen* and *kaikaku* projects in line with MCI target levels: any non-fulfillment was carefully analyzed and immediate corrective measures and actions were carried out to comply with MCPD Principle No. 1.

4.4.2 Step 6: MCI Performance Management

With the help of this step, the annual performance on the achievement of CAMPT has been monitored, both at AMIB (annual manufacturing improvement budget for existing products; multiannual manufacturing improvement budget for new products) level and AMCIB (annual manufacturing cash improvement budget) level, and the degree of employee involvement to achieve the annual MCI targets (in particular through participation in improvement projects and by measuring and monitoring the evolution of losses and waste and their associated costs).

The annual budget of $125 (the last column in Figure 4.1, in thousands of $) was not significantly exceeded.

At the end of the year "N-1," the annual MCI goal of $5,000 was exceeded. Annual MCI was $5,200 (especially as a result of the *kaikaku* project–equipment replacement out of Process 5; see Figure 4.1). This overrun has had an impact on the decrease of the initial annual manufacturing unit cost of 7.6%. Figure 4.3 shows this annual reduction of 7.6%. The initial average annual manufacturing unit cost of $22.4 ($73,000 per 3,264 units) exceeded the 1.5% annual target of decrease requested by the marketing department to support sales, i.e. manufacturing unit cost improvement target of $20.9 ($68,050 per 3,264 unit) and reached a manufacturing unit cost achieved of $20.7 ($67,700 per 3,264 units). Sales unit variations and cost of manufacturing were not significant compared to those planned in year "N-1."

Principle No. 1 of the MCPD system (annual and multiannual manufacturing target profit from MCI does not change at the level of the whole company) was observed in "N-1," namely $37,000.

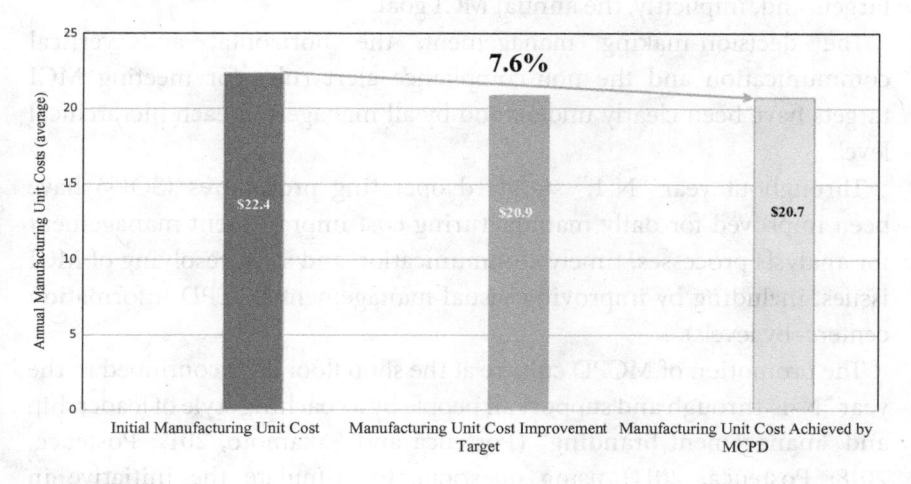

FIGURE 4.3
Annual manufacturing unit cost improvement: initial stage, target and achieved through MCPD system.

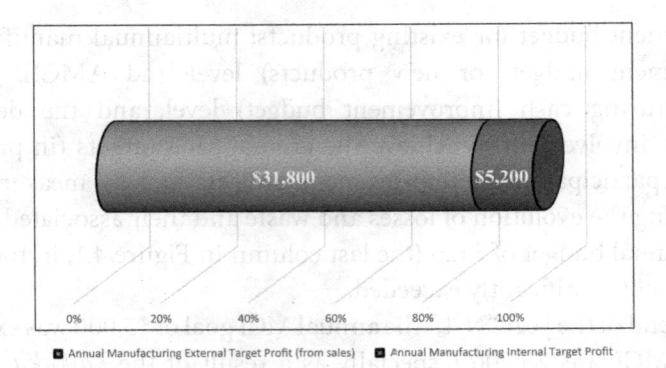

FIGURE 4.4
The full fulfillment of Principle No. 1 of the MCPD system at "BB-Plant" in year "N-1" (annual and multiannual manufacturing target profit from MCI does not change at the level of the whole company).

Figure 4.4 shows the achievement of the annual stake of MCI goal of minimum $5,000.

4.4.3 Step 7: Daily Manufacturing Cost Improvement Management (Act)

With the help of this last step of the MCPD system, the responsibility for performance at each level of "BB-Plant" was achieved to meet annual MCI targets and, implicitly, the annual MCI goal.

The decision-making management, the horizontal and vertical communication and the non-compliance alert rules for meeting MCI targets have been clearly understood by all managers at each hierarchical level.

Throughout year "N-1," standard operating procedures (SOPs) have been improved for daily manufacturing cost improvement management for analysis processes, timely communication and swift resolving of MCI issues, including by improving visual management (MCPD information centers–by levels).

The promotion of MCPD culture at the shop floor level continued in the year "N-1" through and support all people by a coaching style of leadership and "management branding" (Posteucă and Sakamoto, 2017; Posteucă, 2018; Posteucă, 2011) using questions to stimulate the initiative in identifying and implementing MCI means rather than providing solutions at the time. This way to address the CLW KPIs issues and especially the CCLW has increased employees' morale and commitment. All leaders addressed issues directly, peer to peer, and escalated exceptional matters on time.

This approach to management branding (Posteucă and Sakamoto, 2017; Posteucă, 2018; Posteucă, 2011) increases the morale and commitment of employees regarding the application of MCPD system at shop floor level.

4.5 MCPD SYSTEM EFFECTS

Even if the annual MCI target of 5,000 can be considered as a relatively small amount at one year's level, it is absolutely necessary to achieve it, since the health of the entire manufacturing system from the perspective of productivity (capacity, quality and delivery) is controlled this way. Therefore, besides the financial data presented above, using the MCPD system, "BB-Plant" obtained the following operational improvements in the year "N-1":

- Increase from 3,000,000 units in "N-2" to 3,275,000 units in "N-1";
- Increase of the average OEE for the ten strategic pieces of equipment by 3.5%;
- Decrease of the average set-up time by 12.5 minutes;
- Decrease of the man-hours per product by 4.7%;
- Reducing the original scrap ratio percentage by 3.15%;
- Increasing OTIF growth from 94% to 97%;
- Decrease of initial total factory lead time by 12% (measured in days);
- Decrease of initial manufacturing lead time by 15% (measured in days);
- Zero accidents;
- Zero environmental incidents; and last but not least,
- Abandoning premature or unnecessary investment, especially for equipment, by maximizing the capacities of the current equipment (MCPD system focuses on obtaining *internal productivity* rather than obtaining *external productivity* through successive acquisitions of increasingly productive equipment).

4.6 CONCLUSIONS

In conclusion, applying the MCPD system to "BB-Plant" is a strategic and operational managerial success, especially at the shop floor level.

The presentation of CCLW in the profit and loss statement is the way to motivate managers to continuously engage, sustain and employ ongoing efforts to continuously improve productivity in "BB-Plant."

In fact, through the MCPD system, management teams at all levels look at the financial reality mirrored in the reality of processes on the shop floor (Posteucă, 2018, pp. 35–37).

In the absence of the MCPD system at "BB-Plant" in year "N-1," the annual manufacturing target profit of $37,000 was exclusively the task of the annual manufacturing target profit (from sales), and the pressure on the manufacturing system was a creator of losses and waste at the level of processes which was quite difficult to notice, but surely, both in the short term and especially in the medium and long term. So, in the absence of an annual manufacturing internal profit of $5,200, annual manufacturing profit would have been just $31,800 and not necessarily the annual manufacturing target profit of $37,000.

Part II

Profitable Improvement Projects Implementation through MCPD for MCI toward Maximizing Outputs

The second part of this book addresses the implementation of profitable CCLW improvement projects through MCPD to meet MCI in order to consistently support the external manufacturing profit through maximizing outputs (hypostasis of "E" scenario; see Table 2.1). The selection, implementation and monitoring of lucrative projects (or annual MCI means) to achieve the MCI targets is part of step 5 of the MCPD and is usually achieved during the work program, on a continuous basis, alongside the activities to achieve the target production volumes.

Therefore, the systematic (*kaizen*) and systemic (*kaikaku*) improvement projects will be presented in this second part, or other MCI means to achieve MCI targets on time, based on credible and capable assumptions for each MCI strategy for change drivers (CD) of MCI to ensure achievable expectations:

- CD1—effectiveness of current equipment;
- CD2—effectiveness of new profitable products;
- CD3—effectiveness of new equipment.

In the first chapter of the second part of this book, Chapter 5, real or annual improvement MCI means projects are presented (SDI 1–CLW 1; see Matrix 7 from Figure 3.14) to achieve the annual MCI targets (SDI 1–CLW 1; see Matrix 6 from Figure 3.12) convergent to the annual MCI goal (and CAMPT), by improving or eliminating CCLW, in the current order of importance of equipment improvements (quality, performance/speed and availability) with the aim of *increasing equipment effectiveness*. Therefore, Section 5.1 outlines the approach method and results of two real *kaizen* projects to achieve the MCI targets by improving or eliminating CCLW related to equipment quality; one *kaizen* project for equipment scrap reduction and one for equipment rework reduction respectively. Section 5.2 presents two real *kaizen* projects to achieve the MCI targets by improving or eliminating CCLW related to equipment performance (speed); one *kaizen* project for equipment cycle time reduction and one *kaizen* project for people cycle time reduction (eliminating bottleneck). Section 5.3 presents two real *kaizen* projects to achieve the MCI targets by improving or eliminating CCLW related to equipment availability; one *kaizen* project to eliminate equipment breakdown and another *kaizen* project for equipment setup, settings and adjustment time reduction.

The second chapter of this second part of the book, Chapter 6, presents two real *kaikaku* projects or annual MCI means (SDI 1–CLW 2 and CLW3; see Matrix 7 from Figure 3.14 to achieve the annual MCI targets (SDI 1–CLW 2 and CLW 3; see Matrix 6 from Figure 3.12 convergent to the annual MCI goal (and CAMPT), a *kaikaku* project for the development of a new product and a *kaikaku* project for the installation of new equipment respectively, with the aim of *increasing manufacturing capacity through new products and new equipment implementation*.

5

MCI and Increasing Equipment Effectiveness

Continuous improvement of current equipment efficiency is the core concern of the MCI change driver on the need to increase equipment efficiency to ensure external manufacturing target profit.

Achieving the volume of quality target production is the concern of everybody in the company, followed by concern for speed and concern for the availability of equipment.

Continuous knowledge of LW, CLW and CCLW level for equipment and assembly lines and CCLW reduction by planning and running the required annual MCI means at the required deadlines is a must. As with any MCI project, the teams addressing CCLW must determine whether they address the entire losses by the necessary improvement or take only a partial approach, and establish the CCLW structure in detail, i.e. the total CLW of all PFC processes in a CCLW. If it is decided to address the current level of CCLW through several *kaizen* or *kaikaku* projects, then the contribution of each project to the annual MCI goal must be established (through CAMPT).

5.1 MCI AND INCREASING EQUIPMENT QUALITY

Section 5.1.1 presents a *kaizen* project for MCI through equipment scrap reduction, and Section 5.1.2 presents a *kaizen* project for MCI through equipment rework reduction.

5.1.1 Improvement Project 1: Systematic MCI by Scrap Reduction

Company background:

- "CC-Plant":
  - Type of industry: manufacturing and assembly–electrical products;
  - Manufacturing regime: repeated lot–machine shop, transfer lines and assembly lines;
  - PFC number: 1;
  - Improvement area: assembly line 1;
- Company productivity vision (CPV): (1) manufacturing profit increase; (2) number of products sold increase (global market share); (3) manufacturing cost decrease; (4) safety, health, & environment improvement and (5) morale increase.
- From company productivity vision (CPV) through MCI to annual MCI (see Figure 5.1 and Figure 3.1):
  - Annual MCI by 6%;
  - Number of products sold increase by 10%:
    - *Reduce scrap by 15%: equipment scrap losses decrease;*
  - Improve productivity by 8%;
  - Improve Safety, Health and Environment (SHE)–index 10%;
  - Morale increase: increase of the number of improvement projects implemented by 9%.

Note 1: for all the CCLW improvement projects presented below in the next section, the companies have developed CPV and CPM (see Figure 3.1 and Figure 3.2).

Note 2: from the perspective of the MCPD system, which is created on the Plan-Do-Check-Act cycle structure (PDCA cycle), all *kaizen* and *kaikaku* projects for MCI go through the same three phases and seven steps as the entire MCPD system. Therefore, PDCA cycle is cascading and followed within each MCI improvement project with the help of the *catchball process* (Posteucă and Sakamoto, 2017).

5.1.1.1 Phase 1: Manufacturing Cost Policy Analysis

Step 1: From Company Productivity Vision to Establish Annual MCI Goal: Current State of the Equipment from the CCLW (SDI 1–CLW 1) Perspective:

- Annual MCI goal: $5,500,000;
- Number of MCI means (annual strategic improvement projects to meet annual MCI goal by CAMPT): nine *kaizen* projects and one *kaikaku* project (similar to Figure 3.13);
- "CC-Plant" joined the MCPD system three years ago;
- Stake to improve CLW associated with scarp: 17.5% of annual MCI goal (3rd place as weight; three strategic *kaizen* strategic projects were selected out of nine).

Step 2: Target Setting for CCLW: Full Understanding of the Current State of Losses and Waste

- After completing and carefully analyzing the 7 matrices (for PFC analyzed), annual CCLW-related scrap identified in operation number 7 of assembly line 1: $35,000 (see Figure 5.1);
- The theme of the first *kaizen* project for scrap: removing scrap from handling on line 1 in operation 7;

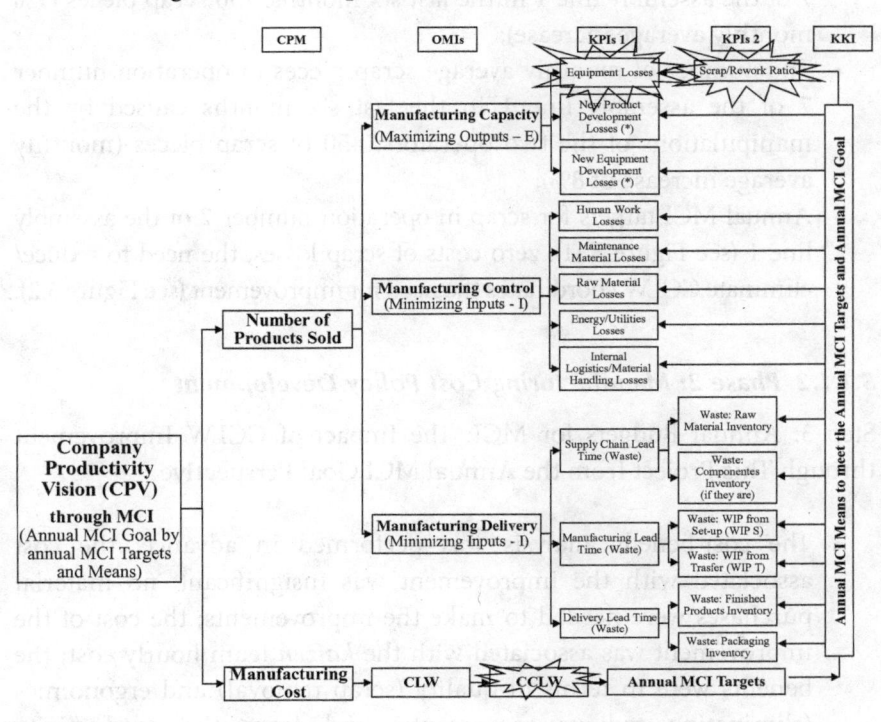

FIGURE 5.1
MCI policy deployment for a PFC for strategically targeting scrap and rework improvement.

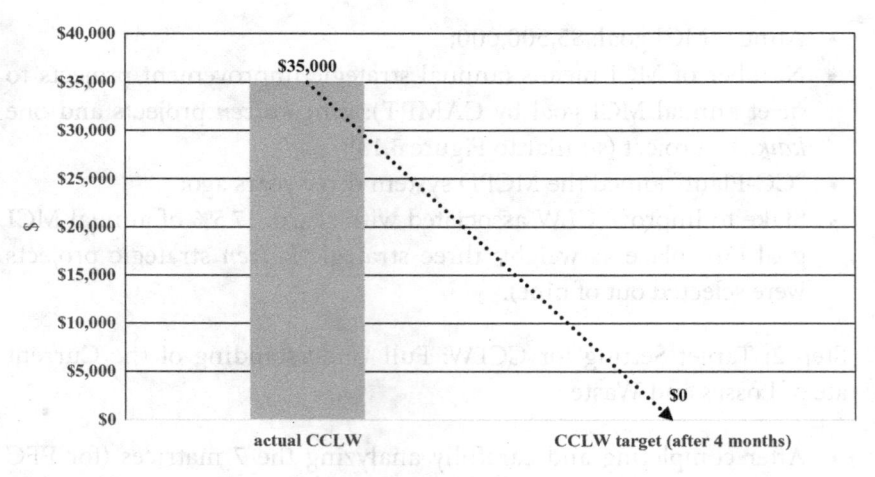

FIGURE 5.2
Setting annual CCLW target for scrap for the assembly line

- The number of monthly average scrap pieces in operation number 7 of the assembly line 1 in the last six months: 750 scrap pieces (7% monthly average increase);
- The number of monthly average scrap pieces in operation number 7 of the assembly line 1 in the last six months caused by the manipulations of the two operators: 450 of scrap pieces (monthly average increase of 8%);
- Annual MCI targets for scrap in operation number 7 of the assembly line 1 (see Figure 3.11): zero costs of scrap losses; the need to reduce/eliminate CCLW coordinates the need for improvement (see Figure 5.2).

5.1.1.2 Phase 2: Manufacturing Cost Policy Development

Step 3: Annual Budgets for MCI: The Impact of CCLW Improvement through This Project from the Annual MCI Goal Perspective

- The cost-benefit analysis was performed in advance; the cost associated with the improvement was insignificant; no material purchases were needed to make the improvements; the cost of the improvement was associated with the *kaizen* team hourly cost; the benefits were in terms of quality (scrap removal) and ergonomics (eliminating tedious movements) and from the productivity perspective (reducing cycle time and direct labor costs to operation 7); the profitability of the lost opportunity was analyzed to achieve production—the equivalence of quality parts produced in exchange for monthly average scrap pieces;

- Cost of physical losses (PL): the cost of the 450 monthly average scrap pieces was \$2,025 per month;
- Cost of total time-related costs (TRL): the total cost on the assembly line and downstream and downstream processes was \$892 per month;
- Total annual CCLW: \$35,000 per year;
- The MCPD project team has carefully followed the manufacturing improvement budgets cycle (the five steps described in Section 3.3.1).

Step 4: Action Plan for MCI: From Setting the CCLW Target to Implementing the New Improvement Standard

- A team of four members and one project leader was established;
- The two-month deadline was set to identify improvements and another two months was allowed for full implementation of improvements: the new standard for the three-day shifts, including the operators learning and training on the new standard operating procedure (SOP);
- The main tasks of the team were:
  - To perform an assessment of the current state of the scrap generation process;
  - To check the current status of the scrap-related CLW recorded in the system over time and to check the current level of the CCLW declared in the system;
  - To prepare a plan of consistent measures to eliminate costs of scrap losses;
  - To analyze scrap generation human errors;
  - To propose countermeasures for identified problems–CLW improvement activities–scrap;
  - To define new standards of work for current work items or by changing the current working method—changing work items–using the method design concept (MDC) (Posteucă and Sakamoto, 2017);
  - To establish the further optimizations required for newly defined standards;
  - To set the frequency of monitoring of the new standards–the list of periodic checks on the new standard.

5.1.1.3 Phase 3: Manufacturing Cost Policy Management

Step 5: Engage the Workforce for MCI: Analyzing and Identifying Improvement Solutions to Reach the CCLW Target

- These have been identified: takt time: 45 seconds; number of operations: 24; number of products assembled on line 1: 6 products;
- Time and motion study has been carried out to analyze the 6 scrap creation items for all 6 products manufactured on line 1; cycle time measured for the operation 7: 42 seconds; cycle time standard for the operation 7: 38 seconds (see Figure 5.3);
- The three landmarks that are assembled in operation 7 were identified: landmark 1: 2 pieces; landmark 2: 1 piece and landmark 3: 2 pieces;
- The tools and devices used for assembly have been identified;
- All types of scrap generating products have been identified;
- The problems of strikes and scratches were identified on the surface of the pieces during operations in item 7 (3 types of scrap causing operations): tower 1 piece "x"; fixing the component "a," turn 2 piece "x" (see Figure 5.3) (Posteucă and Sakamoto, 2017, pp. 327–352);
- Improvement solutions have been identified (see Figure 5.3);
- The new cycle time is 17 seconds.

Step 6: MCI Performance: Checking the CCLW Improvements Results

- The level of CCLW for scrap caused by operators' manipulation has fallen to zero after 3 months since the *kaizen* project started (at line 1–operation 7);
- Annual MCI targets for scrap was met (at line 1–operation 7);
- The MCPD project team has carried out the performance analysis of manufacturing improvement budget targets and the actual manufacturing budgets drawn up; the targets have been met;
- The team has carried out the analysis of the performance of the cash flow improvement budget targets and the actual cash budgets drawn up; the targets have been met;
- The MCPD project team carried out the evaluation of the PFC losses and waste performance indicators after implementing CCLW and CLW improvements (after 30 days, following MCI management process–see Section 3.4.3).

No.	Operating Procedure (work elements)	Standard time	Actual time	Dif.	Cycle Time (seconds) Time Study	Elimination (E)	Combination (C)	Rearrangement (R)	Simplification (S)	Automation (A)	Motion	Walking	Release workload	Hand holding	Inspection	Waiting	Component	Instruments	Assembled piece	Conveyor	Opportunity	Action required	Effects (time – sec.)
1	Retrieve piece "x"	4	4						x												o1	a1	3
2	Turn piece "x"	7	7			x						x									o2	a2	0
3	Retrieving component "a"	6	6																				6
4	Fixing the component "a"		4	4		x								x							o3	a3	0
5	Assembly Component "a"	5	5			x															o4	a4	6
6	Turn piece "x"	9	9			x						x									o5	a5	0
7	Release the piece "x" piece	2	2			x															o6	a6	2

Line name: 1
Date of measurement:
Process: Operation 7

Table header spans: TIME / Cycle Time (seconds) / Time Study; Directions for improvement — Method Design Analysis, Motion study, Materials & Line, Improv.

FIGURE 5.3

Establish opportunities to improve CCLW for scrap at the assembly line.

- The MCPD team leader of this project has conducted an evaluation of the degree of involvement of team members to achieve the MCI target of this project.

Step 7: Daily MCI Management: Standardization and Expansion of Validated Solutions

- The team has developed and implemented a new standard for operation 7 from line 1 to avoid scrap from operations. One component from the subassembly associated to operation 7 was mounted on operation 7 to avoid scrap from operations;
- The MCPD team leader of this project has continuously updated the state of the MCI improvement project at the MCPD system information center.

Future activities:

- Following this *kaizen* for MCI project (cost of scrap losses), a rebalancing of the line was required after the reduction/elimination of scrap generating work elements. At the same time, three new opportunities for improving scrap on line 1 have been identified in other operations.

Conclusion

Therefore, this MCI systematic improvement project, along with other MCI improvement projects, has met the annual MCI goal at "CC-Plant". "CC-Plant" has introduced a series of systemic structural change projects such as: 5S expansion in all areas of the company (including office and warehouse areas), operative maintenance, preventive maintenance, quality maintenance, quality assurance, education and training, redesigning the cost and budget system to capture the variations between the target cost that has been set and the actual cost that has been achieved and the differences between the target cost and the zero cost of losses and waste. "CC-Plant" continuously monitors the price level of PFC-related products, the required unitary profit level, the manufacturing unit cost required to be reached and the CLW and CCLW level for each PFC. "CC-Plant" has a system of prioritization in choosing systematic and systemic strategic annual improvement projects to achieve annual MCI goal. "CC-Plant" has a symbiosis between the traditional budgeting and costing system and the

AMIB (for existing and new products) and AMCIB. The top management at "CC-Plant" keeps track of what is the normal cost reduction potential for each PFC and each product by continually measuring CLW, by highlighting CLW and CCLW in company's internal financial statements and by strategic planning of annual MCI means. In this way, top management consistently supports vision, mission and strategies of productivity in the short term and especially in the medium and long term. Finally, daily MCI management ensures that every employee is committed to achieving the annual MCI goal and quickly and consistently solves any cost problem, and not only, on the basis of an effective and efficient dialogue between all levels of the company.

5.1.2 Improvement Project 2: Systematic MCI by Rework Reduction

Company background:

- "DD-Plant" joined the MCPD system three years ago;
- Type of industry: manufacturing and assembly–parts for automotive industry;
- Manufacturing regime: repeated lot–machine shop, transfer lines and assembly lines;
- PFC number: 3;
- Improvement area: the painting line 1.

5.1.2.1 Phase 1: Manufacturing Cost Policy Analysis

Step 1: From Company Productivity Vision to Establish Annual MCI Goal: Current State of the Process/Equipment from the CCLW Perspective

- Annual MCI goal: $4,750,000;
- Number of MCI means (annual strategic improvement projects to meet annual MCI goal by CAMPT): 12 *kaizen* projects and 2 *kaikaku* project (similar to Figure 3.13);
- Stake to improve CLW associated with rework: 27.5% of annual MCI goal (3rd place as weight; 2 strategic *kaizen* strategic projects were selected out of 12).

Step 2: Target setting for CCLW: Full Understanding of the Current State of Losses and Waste

- After completing and carefully analyzing the 7 matrices (for PFC analyzed), annual CCLW related rework identified at the painting line: $75,000 (see Figure 5.1);
- The theme of the first *kaizen* project for rework: removing rework for injected plastic parts with impurities;
- Information about the automatic painting installation: cycle time is 16 seconds per painted piece; production per shift is 1,420 painted pieces and OEE is 85%;
- Analysis of non-compliant landmarks: the six months average of the non-compliant painted landmarks from the compliant painted landmarks was 23.7%;
- The preliminary analysis of the causes of non-compliant landmarks: (1) impurities on the commercial surface; (2) plastic injection mold defects; (3) insufficient paint; (4) paint splashes;
- The main cause of the non-compliant landmarks: impurities on the commercial surface in the proportion of 18.6% of the total painted parts;
- Annual CCLW for rework of plastic landmarks injected with impurities on the commercial surface from the paint line: $60,000;
- Annual MCI targets for rework (see Figure 3.11): zero costs of rework losses; the need to reduce/eliminate CCLW coordinates the need for improvement (see Figure 5.4).

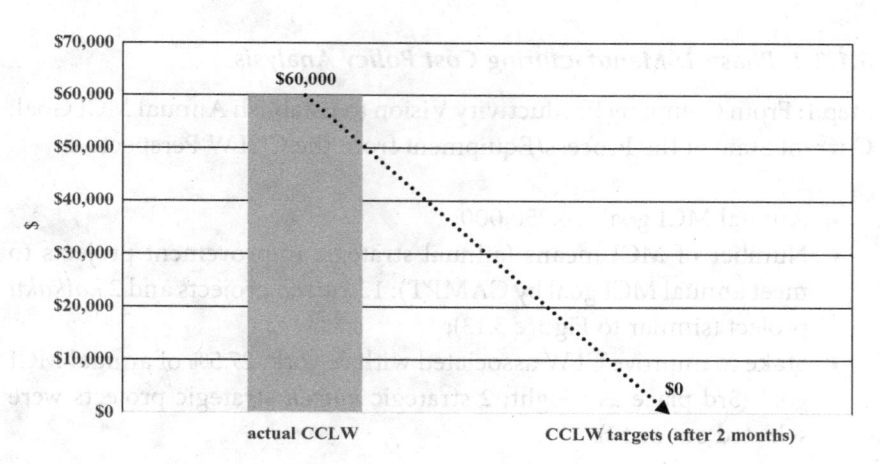

FIGURE 5.4
Setting annual CCLW target for eliminating rework of painted plastic parts on the paint line.

5.1.2.2 Phase 2: Manufacturing Cost Policy Development

Step 3: Annual Budgets for MCI: The Impact of CCLW Improvement through This Project from the Annual MCI Goal Perspective

- The cost-benefit analysis was performed in advance; the cost associated with the improvement was insignificant; the costs of improvement were mainly associated with the *kaizen* team's hourly cost;
- The profitability of the lost opportunity to achieve production was assessed–the equivalent of quality parts manufactured in exchange for rework parts;
- Cost of physical losses (PL): the cost of materials for the 264 rework parts per shift or 23,770 per month was $4,045 per month or $48,550 per year;
- Cost of total time-related losses (TRL): the total cost over the entire painting line and upstream and downstream processes was $954 per month or $11,450 per year;
- Total annual CCLW: $60,000 per annum;
- The MCPD project team has carefully followed the manufacturing improvement budgets cycle (the five steps described in Section 3.3.1).

Step 4: Action Plan for MCI: from Setting the CCLW Target to Implementing the New Improvement Standard

- A team consisting of six members and a project leader was established;
- The three-week deadline was set to identify improvements and another five weeks for full implementation of the improvements (the new standard);
- The main tasks of the team were:
  - To perform an assessment of the current state of the rework generating process of plastic landmarks injected with impurities on the commercial surface from the painting line;
  - To check the current status of the CLW associated with this type of rework recorded in the system over time and to check the current level of the CCLW declared in the system;
  - To prepare a plan of consistent measures to eliminate costs of rework losses;
  - To propose countermeasures for identified problems;
  - To define new standards of work for current work items;

- To establish the further optimizations required for newly defined standards;
- To set the frequency of monitoring of the new standards—the list of periodic checks on the new standard;
- To extend the validated solutions to other painting equipment.

5.1.2.3 Phase 3: Manufacturing Cost Policy Management

Step 5: Engage the Workforce for MCI: Analyzing and Identifying Improvement Solutions to Reach the CCLW Target

- The team analyzed the technological flow of the painting plant and six technological areas were defined:
  - Loading of plastic pieces for painting;
  - Painting booth;
  - Transfer to the polymerization furnace;
  - Polymerization furnace;
  - Unloading the painted pieces;
  - Air supply of the entire ventilation installation (polymerization furnace and painting booth).
- The impurity control point was analyzed before the transfer to the manufacturing lines;
- The current and standard parameters of the painting equipment have been checked and analyzed to understand its operating principles: conveyor belt speed, paint pump pressure, air spray pressure, polymerization temperature and temperature inside the painting booth; all the measured parameters were in their standard values;
- All types of paint nonconformities have been defined for rework for injected plastic pieces with impurities;
- Rework generating operations have been identified and analyzed: (1) transport of plastic landmarks for rework; (2) landmark surface polishing; (3) grinding of polished pieces with a piece of textile material imbided in isopropyl alcohol; (4) transport of plastic landmarks and reintroduction of the landmarks in the painting process.
- The technological flow of the painted injected landmark was analyzed in detail;
- All the technological flow operations that are the roots of rework from the perspective of: man, machine, material and method (4M) have been identified, analyzed and evaluated;

- A plan of measures, responsibilities and deadlines for the problems identified has been established;
- Improvement solutions have been identified: (1) changing the shape and size of the devices on which the landmarks are positioned for painting; (2) changing the "5S" standard for cleaning—reducing the cleaning range of the conveyor from once a day to once per shift; (3) blowing the injected landmarks with ionized air before loading the painting plant; (4) setting up a sealed enclosure for the rework operation of the painted injected landmarks (location available; no additional fittings required).

Step 6: MCI Performance: Checking the CCLW Improvements Results

- The level of CCLW for rework of plastic pieces injected with impurities on the commercial surface from the paint line has reached zero within two months of implementing the solutions;
- Annual MCI targets for rework of plastic pieces injected with impurities has been fully met;
- The MCPD project team has carried out the performance analysis of manufacturing improvement budgets targets and actual manufacturing budgets drawn up; the targets have been met;
- The team has carried out the analysis of the performance of the cash flow improvement budget targets and the actual cash budgets drawn up; the targets have been met;
- The MCPD project team carried out the evaluation of the PFC losses and waste performance indicators after implementing CCLW and CLW improvements (after 30 days, following MCI management process–see Section 3.4.3);
- The MCPD team leader of this project has conducted an evaluation of the degree of involvement of team members to achieve the MCI target of this project.

Step 7: Daily MCI Management: Standardization and Expansion of Validated Solutions

- The team has developed and implemented a new standard for the specific working instructions for the painting equipment to avoid rework due to impurities;

- The MCPD team leader of this project has continuously updated the state of the MCI improvement project at the MCPD system information center.

Future activities:

- Another project *kaizen* for MCI for rework of plastic pieces injected was planned to reduce the remaining 5.1% of remaining rework types: (1) plastic injection mold defects; (2) insufficient paint; and (3) paint splashes.

Conclusion

Therefore, this MCI systematic improvement project, along with other MCI improvement projects, has met the annual MCI goal at "DD-Plant". "DD-Plant" continuously monitors the price level of PFC-related products, the required unitary profit level, the manufacturing unit cost required to be reached and the CLW and CCLW level for each PFC. "DD-Plant" has a symbiosis between the traditional budgeting and costing system and the AMIB (for existing and new products) and AMCIB. The top management at "DD-Plant" keeps track of what is the normal cost reduction potential for each PFC and each product. Finally, daily MCI management ensures the stability of the manufacturing system and the continued growth of people's trust that they are on the right track in achieving production planning and improvements.

5.2 MCI AND INCREASING MANUFACTURING PROCESS CAPACITY BY ELIMINATING EQUIPMENT BOTTLENECK

Section 5.2.1 presents a *kaizen* project for MCI through equipment cycle time reduction and Section 5.2.1 presents a *kaizen* project for MCI by reducing the cycle times of operations of a manual assembly line.

5.2.1 Improvement Project 3: Systematic MCI by Equipment Cycle Time Reduction

Company background:

- "EE-Plant" joined the MCPD system two years ago;

- Type of industry: manufacturing and assembly–metal products;
- Manufacturing regime: repeated lot–machine shop, press lines and assembly lines;
- PFC number: two;
- Improvement area: machine shop.

5.2.1.1 Phase 1: Manufacturing Cost Policy Analysis

Step 1: From Company Productivity Vision to Establish Annual MCI Goal: Current State of the Process/Equipment from the CCLW Perspective

- Annual MCI goal: $3,500,000;
- Number of MCI means (annual strategic improvement projects to meet annual MCI goal by CAMPT): eight *kaizen* projects and one *kaikaku* project (similar to Figure 3.13);
- Stake to improve CLW associated with equipment cycle time reduction: 25% of annual MCI goal (1st place as weight; one *kaizen* strategic project was selected out of eight).

Step 2: Target setting for CCLW: Full Understanding of the Current State of Losses and Waste

- After completing and carefully analyzing the 7 matrices (for PFC analyzed), annual CCLW related with equipment cycle time reduction: $875,000;
- The theme of the *kaizen* project for equipment cycle time reduction: capacity increase for "ZZ" equipment;
- Information on the "ZZ" equipment:
  - Process: pre-assemble to the product "AA";
  - Daily work schedule: three shifts, 1,315 minutes;
  - Daily production capacity: 2,160 pieces (takt time: 31 seconds; OEE: 85%)
  - Required daily production capacity: 2,678 pieces (takt time required: 25 seconds; OEE: 85%; including meeting the annual CCLW target of $875,000);
  - The initial cycle time standard was 25 seconds;
- Annual MCI targets for equipment cycle time reduction (see Figure 3.11): zero costs of speed losses; the need to reduce/eliminate CCLW coordinates with the need for improvement (see Figure 5.5);

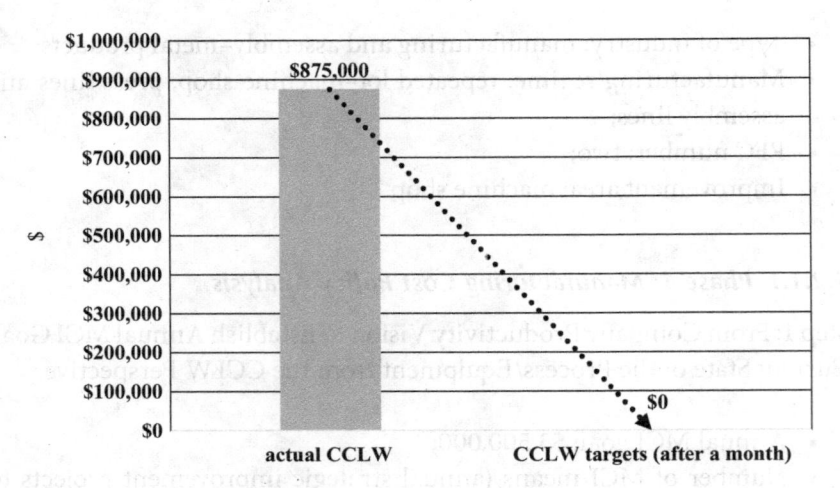

FIGURE 5.5
Setting annual CCLW target for equipment cycle time reduction

- Annual MCI means targets for equipment cycle time reduction (see Figure 5.6 and Figure 5.7).

5.2.1.2 Phase 2: Manufacturing Cost Policy Development

Step 3: Annual Budgets for MCI: The Impact of CCLW Improvement through This Project from the Annual MCI Goal Perspective

- The cost-benefit analysis was carried out in advance; the cost associated with the improvement was insignificant; the costs of improvement were mainly associated with the *kaizen* team's hourly cost;

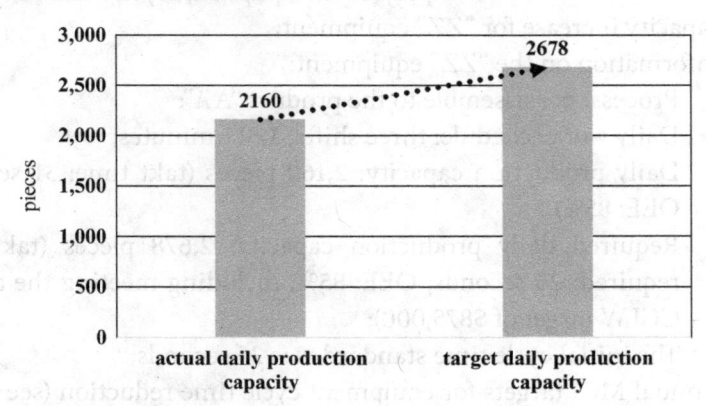

FIGURE 5.6
Set the target for increasing the daily capacity of the "ZZ" equipment

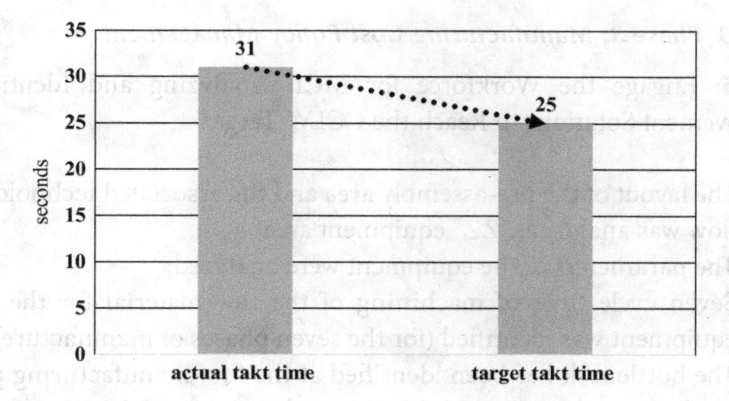

FIGURE 5.7
Set the target of reducing the cycle time to "ZZ" equipment to meet the new takt time

- The profitability of the lost opportunity to achieve production was assessed–the equivalent of the pieces for equipment cycle time reduction;
- The MCPD project team has carefully followed the manufacturing improvement budgets cycle (the five steps described in Section 3.3.1).

Step 4: Action Plan for MCI: from Setting the CCLW Target to Implementing the New Improvement Standard

- A team consisting of three members and one project leader was established;
- The four-week deadline has been set to identify improvements and to fully implement them (the new standard);
- The main tasks of the team were:
  - To carry out an assessment of the current status of the "ZZ" equipment;
  - To check the current status of the CLW associated to equipment speed losses and check the current level of the CCLW declared in the system;
  - To prepare a plan of consistent measures to eliminate costs of speed losses;
  - To propose countermeasures for problems identified at the equipment cycle time level;
  - To define new standards of work for the current work cycle equipment;
  - To extend validated solutions to other similar equipment.

5.2.1.3 Phase 3: Manufacturing Cost Policy Management

Step 5: Engage the Workforce for MCI: Analyzing and Identifying Improvement Solutions to Reach the CCLW Target

- The layout of the pre-assembly area and the associated technological flow was analyzed: "ZZ" equipment area;
- The parameters of the equipment were analyzed;
- Seven cycle time of machining of the raw material for the "ZZ" equipment was identified (for the seven phases of manufacture);
- The bottleneck has been identified at the first manufacturing phase (cycle time of 31 seconds): loading with raw materials;
- The other six operations had a maximum cycle time of 21 seconds;
- A bottleneck cycle time analysis was performed to identify improvement opportunities: loading with raw materials (quite similar to Figure 5.3);
- Twelve elements of work were identified in the first manufacturing phase; three of these were waiting and were eliminated; four of them have cut times by simplifying how to work without investment;
- The new cycle time of loading with raw materials was lowered to 23 seconds.

Step 6: MCI Performance: Checking the CCLW Improvements Results

- CCLW level for equipment cycle time reduction reached 23 seconds in four weeks since the *kaizen* project for MCI started;
- Annual MCI targets for equipment cycle time reduction was fully met ($875,000)
- The new possible takt time is 23 seconds;
- New daily production capacity: 2,911 pieces (the initial target was 2,678 units);
- The MCPD project team has carried out the performance analysis of manufacturing improvement budget targets and the actual manufacturing budgets drawn up; the targets have been met;
- The team has carried out the analysis of the performance of the cash flow improvement budget targets and the actual cash budgets drawn up; the targets have been met;
- The MCPD project team carried out the evaluation of the PFC losses and waste performance indicators after implementing CCLW and

CLW improvements (after 30 days, following MCI management process–see Section 3.4.3);

- The MCPD team leader of this project has conducted an evaluation of the degree of involvement of team members to achieve the MCI target of this project.

Step 7: Daily MCI Management: Standardization and Expansion of Validated Solutions

- The team has developed and implemented a new standard for "ZZ" equipment cycle time;
- The MCPD team leader of this project has continuously updated the state of the MCI improvement project at the MCPD system information center.

Future activities:

- The improvement expansion to other similar equipment has been planned (loading with raw materials).

Conclusion

Therefore, this MCI systematic improvement project, along with other MCI improvement projects, has met the annual MCI goal at EE-Plant. EE-Plant has introduced a series of systemic structural change projects such as: 5S expansion in all areas of the company, operative maintenance, preventive maintenance, quality assurance, education and training, redesigning the cost and budgeting system.

5.2.2 Improvement Project 4: Systematic MCI by Cycle Time Reduction at an Assembly Line

Company background:

- "FF-Plant" joined the MCPD system two years ago;
- Type of industry: manufacturing and assembly; household appliances;
- Manufacturing regime: repeated lot–machine shop, plastic injection and assembly lines;
- PFC number: 4 (4 assembly lines);
- Improvement area: assembly lines.

5.2.2.1 Phase 1: Manufacturing Cost Policy Analysis

Step 1: From Company Productivity Vision to Establish Annual MCI Goal: Current State of the Process/Equipment from the CCLW Perspective

- Annual MCI goal: $2,500,000;
- Number of MCI means (annual strategic improvement projects to meet annual MCI goal by CAMPT): seven *kaizen* projects and two *kaikaku* project (similar to Figure 3.13);
- Stake to improve CLW associated with people cycle time reduction at an assembly lines: 25% of annual MCI goal (third place as weight; two *kaizen* strategic projects were selected out of the seven to eliminate bottleneck on the line; one of the *kaizen* projects for bottleneck removal was on line 2).

Step 2: Target setting for CCLW: Full understanding of the Current State of Losses and Waste

- After completing and carefully analyzing the 7 matrices (for PFC analyzed), annual CCLW related with bottleneck to assembly line 2: $325,000;
- The theme of the *kaizen* project for bottleneck to assembly line: non-value-added operation reduction at assembly line 2 and fitting each cycle time into the new takt time;
  - Process: assembly line 2;
  - Daily work schedule: 3 shifts, 1,315 minutes;
  - Required daily production capacity: 1,676 pieces (takt time required: 40 seconds; including meeting the annual CCLW target of $325,000);
  - 41 operations in assembly line 2;
- Annual MCI targets for bottleneck to assembly line 2 (see Figure 3.11 and Figure 3.12): zero costs of speed losses; the need to reduce/eliminate CCLW coordinates the need for improvement (see Figure 5.8);
- Annual MCI means targets for bottleneck to assembly line 2: there have been identified five bottleneck operations on assembly line 2 (see Figure 5.9).

5.2.2.2 Phase 2: Manufacturing Cost Policy Development

Step 3: Annual Budgets for MCI: The Impact of CCLW Improvement Through This Project from the Annual MCI Goal Perspective

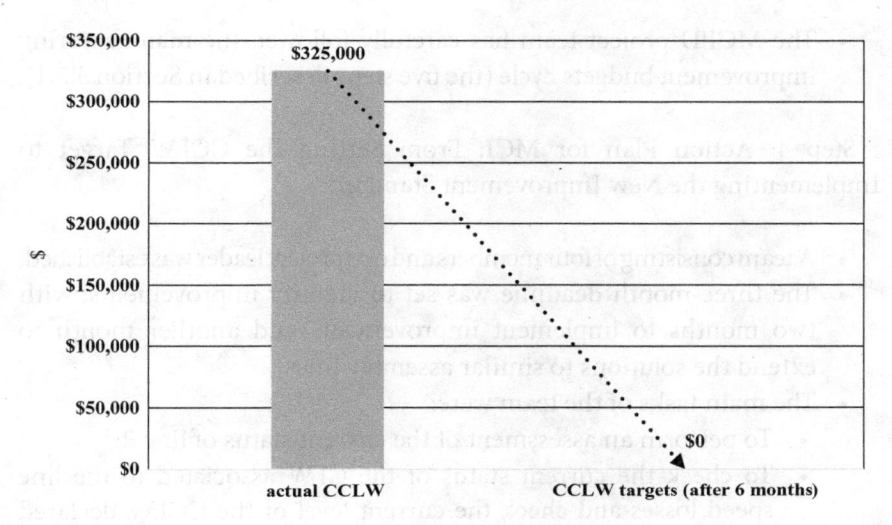

FIGURE 5.8
Setting annual CCLW target for non-value-added operation at the assembly line (for bottlenecks to assembly line)

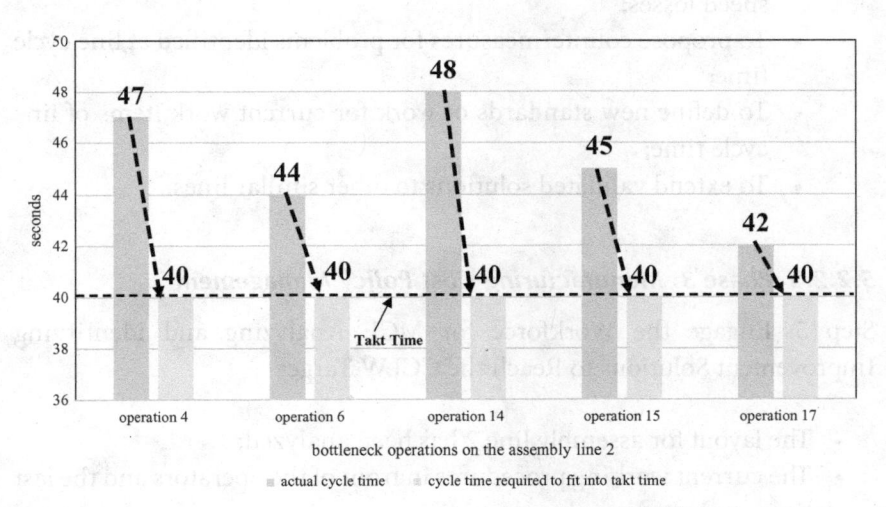

FIGURE 5.9
Set the target of reducing the cycle time on assembly line 2 to meet the new takt time.

- The cost-benefit analysis was performed in advance; the cost associated with the improvement was insignificant; improvement costs amounted to $7,500 for physical layout of line 2 and the costs associated with the *kaizen* team's hourly cost;
- The profitability of the lost opportunity to achieve production was assessed–the equivalent of the pieces for bottlenecks to assembly line 2;

- The MCPD project team has carefully followed the manufacturing improvement budgets cycle (the five steps described in Section 3.3.1).

Step 4: Action Plan for MCI: From Setting the CCLW Target to Implementing the New Improvement Standard

- A team consisting of four members and one project leader was established;
- The three-month deadline was set to identify improvements, with two months to implement improvements and another month to extend the solutions to similar assembly lines;
- The main tasks of the team were:
  - To perform an assessment of the current status of line 2;
  - To check the current status of the CLW associated to the line speed losses and check the current level of the CCLW declared in the system;
  - To prepare a plan of consistent measures to eliminate costs of line speed losses;
  - To propose countermeasures for problems identified at line cycle time;
  - To define new standards of work for current work items of line cycle time;
  - To extend validated solutions to other similar lines.

5.2.2.3 Phase 3: Manufacturing Cost Policy Management

Step 5: Engage the Workforce for MCI: Analyzing and Identifying Improvement Solutions to Reach the CCLW Target

- The layout for assembly line 2 has been analyzed;
- The current working procedures in front of the operators and the last time study were analyzed;
- The necessary level of the takt time has been checked—in line with the necessary capacity planning of the line for the next period;
- Bottleneck operations have been identified: 4, 6, 14, 15 and 17 (see Figure 5.9);
- All the main types of products made on line 2 were identified: six main product types;
- Successive measurements were performed to verify the actual cycle times (including additional work elements);

- All components that are assembled in bottleneck operations were identified;
- All the assembling devices and tools in the bottleneck operations have been identified;
- For each bottleneck operation, all value-added and all non-value-added tasks have been identified and defined for all work elements as ways to reach the CCLW target; for example: the new cycle time for operation 14 is 36 seconds (a drop of 45 seconds, reducing non-value-added work elements); (see Figure 5.10; example for operation 14) (Posteucă and Sakamoto, 2017, pp. 327–352);
- Similarly to operation 14, operations 4, 6, 15 and 17 have been approached to reduce the cycle times and fit them into takt time;
- The main direction of the improvements was focused on the unnecessary movements of the operators;
- The proposed and implemented improvements were:
  - For operation 4: mass redesign for component storage (reduction of 5 meters between processes)
  - For operation 6: placing the roller conveyor for pallet storage (reduction of 1 meter between process);
  - For operation 14: placing the roller conveyor for pallet storage (reduction of 2 meters between process);
  - For operation 15: location of "flow rack" for stowage storage (reduction of 2 meters between process);
  - For operation 17: reduction of distances between operations 17 to 18 (reduction of 0.95 meters between process).

Step 6: MCI Performance: Checking the CCLW Improvements Results

- The level of CCLW for non-value-added operation at the assembly line (for bottlenecks to assembly line) has reached zero after five months since the initiation of the *kaizen* project for MCI (the initial deadline was observed);
- Annual MCI targets for non-value-added operation reduction at the assembly line was fully met ($325,000)
- The new possible takt time is of 40 seconds;
- new daily production capacity: 1,676 pieces (the target for assembly line capacity has been reached);
- The MCPD project team has carried out the performance analysis of manufacturing improvement budget targets and the actual manufacturing budgets drawn up; the targets have been met;

Line name: coffee maker assembly line 2	Date of measurement: day/month/year	Process: Operation 14

TIME — Cycle Time (seconds) — Study of times

No. of work elements	Operating Procedure	value-added: preparation	value-added: type of operation: manual	value-added: type of operation: automatic	non-value-added: move	non-value-added: handling	non-value-added: waiting	non-value-added: positioning	non-value-added: other operations	Total	1	2	3	4	5	6	7	8	9	10	11	12	13	14	15	16	17	18	19	20	21	22	23	24	25	26	27	28	29	30	31	32	33	34	35	36	37	38	39	40	41	42	43	44	45
1	reach and grapple lower shock absorber X				5	2				7																																													
2	move to the J carcass				3					3																																													
3	fixing the lower damper X	3								3																																													
4	four-point screwdriving of the lower damper X		10							10																																													
5	move to the upper damper Y				4					4																																													
6	fixing the upper damper Y	4								4																																													
7	four-point screwdriving of the upper damper Y		8							8																																													
8	deburring hole for passage hose "s"								4	4																																													
9	hose passage "s"		2							2																																													

Directions for improvement

No. of work elements	Method Design Analysis: Elimination (E)	Combination (C)	Rearrangement (R	Simplification (S)	Automation (A)	Motion study: Motion	Walking	Release workload	Hand holding	Inspection	Waiting	Materials & Line: Component	Instruments	Assembled piece	Conveyor	Improv.: Opportunity	Action required	Efects (time – sec.)
1				x		x			x			x						5
2				x			x											2
3																		3
4																		10
5				x			x											2
6																		4
7																		8
8	x								x					x				0
9																		2

FIGURE 5.10
Establish opportunities to improve CCLW for assembly line 2–operation 14.

- The team has carried out the analysis of the performance of the cash flow improvement budget targets and the actual cash budgets drawn up; the targets have been met;
- The MCPD project team carried out the evaluation of the PFC losses and waste performance indicators after implementing CCLW and CLW improvements (after 30 days, following MCI management process–see Section 3.4.3);
- The MCPD team leader of this project has conducted an evaluation of the degree of involvement of team members to achieve the MCI target of this project.

Step 7: Daily MCI Management: Standardization and Expansion of Validated Solutions

- The team has developed and implemented a new SOP for operations 4, 6, 14, 15 and 17;
- The MCPD team leader of this project has continuously updated the state of the MCI improvement project at the MCPD system information center.

Future activities:

- Extensions of improvements from line 2 to line 4 have been planned (line 4 is similar to line 2);
- The team started planning a pilot project on line 2 for redesigning working methods with the help of Methods Design Concept (MDC) (Posteucă and Sakamoto, 2017).

Conclusion

Therefore, this MCI systematic improvement project, along with other MCI improvement projects, has met the annual MCI goal at "FF-Plant". "FF-Plant" has introduced a series of systemic structural change projects such as: 5S expansion in all areas of the company, operative maintenance, preventive maintenance, quality assurance, education and training, redesigning the cost and budgeting system. "FF-Plant" continuously monitors the price level of PFC-related products, the required unitary profit level, the manufacturing unit cost required to be reached and the CLW and CCLW level for each PFC. "FF-Plant" has a system of prioritization in choosing systematic and systemic strategic annual improvement projects to achieve annual MCI goal. "FF-Plant" has a symbiosis between the traditional budgeting and costing system and the AMIB (for existing and new products) and AMCIB. The top management at "FF-Plant" keeps track of what is the normal cost reduction potential for each PFC and each product by continually measuring CLW, by highlighting CLW and CCLW in company internal financial statements and by strategic planning of annual MCI means. In this way, top management consistently supports vision, mission and strategies of productivity in the short term and especially in the medium and long term. Finally, daily MCI management ensures short-run controls that favor rapid and accurate responses to any deviation from the annual MCI goal (by using a good problem-solving process).

5.3 MCI AND INCREASING MANUFACTURING PROCESS CAPACITY BY IMPROVING THE AVAILABILITY OF THE EQUIPMENT

Section 5.3.1 presents a *kaizen* project for MCI through equipment breakdown losses elimination (time with mechanical failure) and Section 5.3.2 presents a *kaizen* project for MCI through equipment set-up, settings, and adjustment losses improvement.

5.3.1 Improvement Project 5: Systematic MCI by Equipment Breakdown Time Elimination

Company background:

- "FF-Plant" joined the MCPD system two years ago;
- Type of industry: manufacturing and assembly; household appliances;
- Manufacturing regime: repeated lot–machine shop, plastic injection and assembly lines;
- PFC number: 4 (four assembly lines);
- Improvement area: horizontal plastic injection molding machine ("AS").

5.3.1.1 Phase 1: Manufacturing Cost Policy Analysis

Step 1: From Company Productivity Vision to Establish Annual MCI Goal: Current State of the Equipment from the CCLW Perspective

- Annual MCI goal: $2,500,000;
- Number of MCI means (annual strategic improvement projects to meet annual MCI goal by CAMPT): seven *kaizen* projects and two *kaikaku* project (similar to Figure 3.13);
- Stake to improve CLW associated with breakdown time elimination: 35% of annual MCI goal (first place as weight; one *kaizen* strategic project was selected out of seven).

Step 2: Target setting for CCLW: Full Understanding of the Current State of Losses and Waste

- After completing and carefully analyzing the 7 matrices (for PFC analyzed), annual CCLW related with mechanical equipment breakdown for horizontal plastic injection molding machine: $875,000;
- The theme of the *kaizen* project for mechanical equipment breakdown: elimination breakdown for injection molds (for "AS"):
  - Process: injection of plastic parts–line 2;
  - Cycle time: 65 seconds;
  - OEE: 83%
  - Daily work schedule: 3 shifts, 1,315 minutes;
  - Daily production (3 shifts): 1,007 parts;
  - OEE target: 84%.
- Annual breakdown types for horizontal plastic injection molding machine (see Figure 5.11):

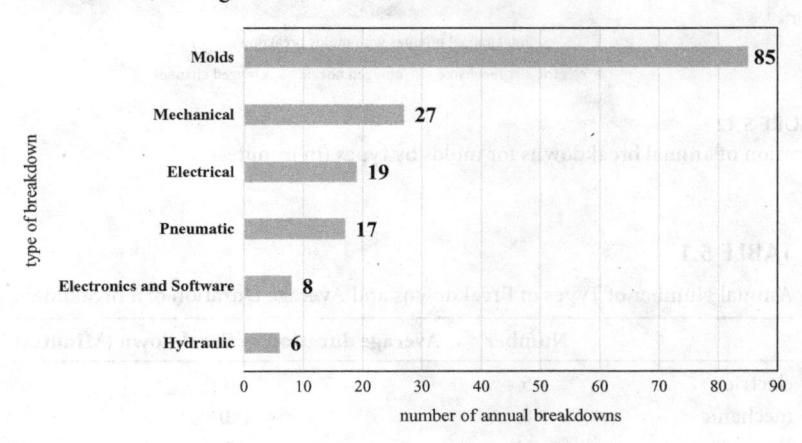

FIGURE 5.11
Annual breakdown types for horizontal plastic injection molding machine.

- Duration of annual breakdowns for molds by types (see Figure 5.12):
- Annual number of types of breakdowns and average duration of a breakdown (see Table 5.1):
- Annual MCI means targets for eliminate clogged nozzle and clogged channel breakdowns (85% of the total breakdowns for molds; see Figure 5.13):
- Annual MCI targets for mechanical equipment breakdown for horizontal plastic injection molding machine (target to eliminate clogged nozzle and clogged channel breakdowns): zero costs of equipment breakdown losses (clogged nozzle and clogged channel breakdowns); the need to reduce/eliminate CCLW coordinates with the need for improvement: $875,000 (see Figure 5.14).

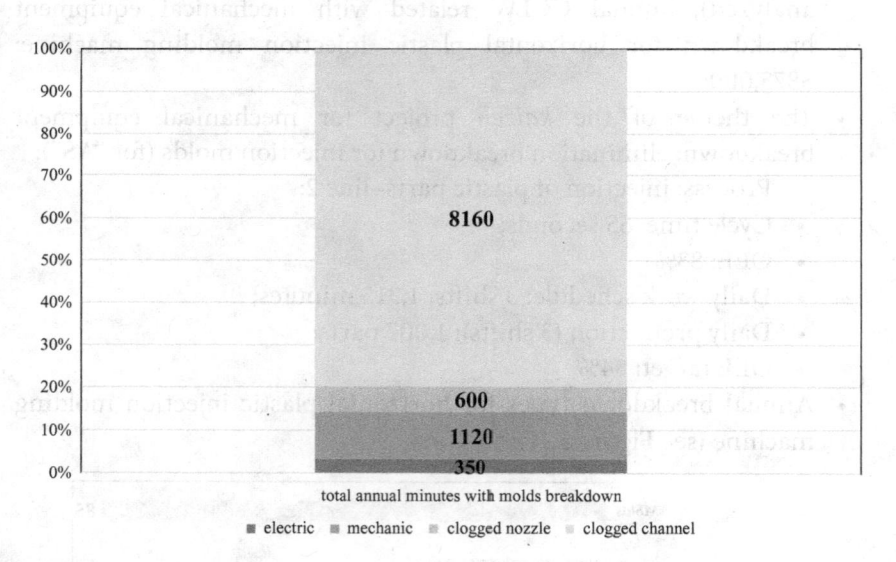

FIGURE 5.12

Duration of annual breakdowns for molds by types (in minutes)

TABLE 5.1

Annual Number of Types of Breakdowns and Average Duration of a Breakdown

	Number	Average duration of Breakdown (Minutes)
electric	5	70
mechanic	8	140
clogged nozzle	24	25
clogged channel	48	170

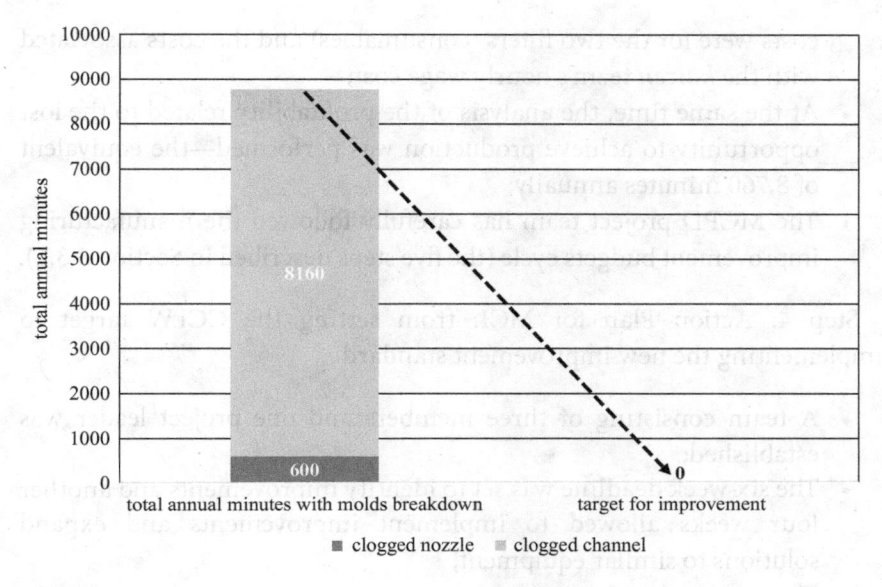

FIGURE 5.13
Target to eliminate clogged nozzle and clogged channel breakdowns.

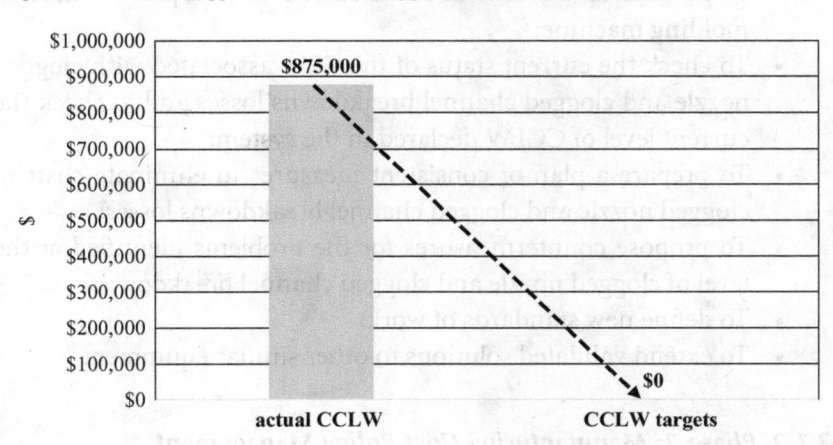

FIGURE 5.14
Setting annual CCLW target for clogged nozzle and clogged channel breakdowns.

5.3.1.2 Phase 2: Manufacturing Cost Policy Development

Step 3: Annual Budgets for MCI: The Impact of CCLW Improvement Through This Project From the Annual MCI Goal Perspective

- The cost-benefit analysis was performed in advance; the cost associated with the improvement was insignificant; the improvement costs were for the two filters (consumables) and the costs associated with the *kaizen* team's hourly wage cost;

- At the same time, the analysis of the profitability related to the lost opportunity to achieve production was performed—the equivalent of 8,760 minutes annually;
- The MCPD project team has carefully followed the manufacturing improvement budgets cycle (the five steps described in Section 3.3.1).

Step 4: Action Plan for MCI: from setting the CCLW target to implementing the new improvement standard

- A team consisting of three members and one project leader was established;
- The six-week deadline was set to identify improvements and another four weeks allowed to implement improvements and expand solutions to similar equipment;
- The main tasks of the team were:
  - To perform an assessment of the current state of plastic injection molding machine;
  - To check the current status of the CLW associated with clogged nozzle and clogged channel breakdowns losses and to check the current level of CCLW declared in the system;
  - To prepare a plan of consistent measures to eliminate costs of clogged nozzle and clogged channel breakdowns losses;
  - To propose countermeasures for the problems identified at the level of clogged nozzle and clogged channel breakdowns;
  - To define new standards of work;
  - To extend validated solutions to other similar equipment.

5.3.1.3 Phase 3: Manufacturing Cost Policy Management

Step 5: Engage the Workforce for MCI: Analyzing and Identifying Improvement Solutions to Reach the CCLW Target

- The standard and current parameters were checked: auger temperature, injection nozzle temperature, mold temperature, auger pressure and closing force; there were no variations against the standard values;

- The 65-second standard cycle time was checked; there were no variations;
- The principles of manifestation of the clogged nozzle and clogged channel breakdowns have been analyzed:
  - (a) incomplete injection of the parts;
  - (b) increase in pressure at injection time on the auger—automatic shutdown of the injection process;
- The maintenance operations have been analyzed in detail for: (a) 170 minutes (clogged channel–clogged mold network) and (b) 25 minutes (clogged nozzle);
- The root cause analysis for clogged nozzle and clogged channel breakdowns has been performed; the root cause was: impurities in molten material; the operation method was observed and the mold had a conforming design;
- Necessary improvements resulting from analysis:
  - (1) implementing a ferrous impurities filtering system;
  - (2) filtering on the injection molding nozzle of the molten material (filtration system of the melted material prior to injection molding).

Step 6: MCI Performance: Checking the CCLW Improvements Results

- Breakdown types for horizontal plastic injection molding machine after six months of implementation of improvements has been reduced (see Figure 5.15); all 18 new reported incidents were caused by clogged channel–clogged mold network:
- The level of CCLW target for breakdowns of horizontal plastic injection molding machine (molds: clogged nozzle and clogged channel) was reached at the end of the year following the *kaizen* project for MCI (the initial ten-week deadline was observed);
- Annual MCI targets for breakdowns of horizontal plastic injection molding machine ("AS") has been fully met ($875,000); the sum of the CLW results for the entire PFC and the system due to the CLW of the breakdowns (including costs of waste and cost of scrap resulting following breakdowns);
- The level of OEE increased to 84%; the target has been reached;

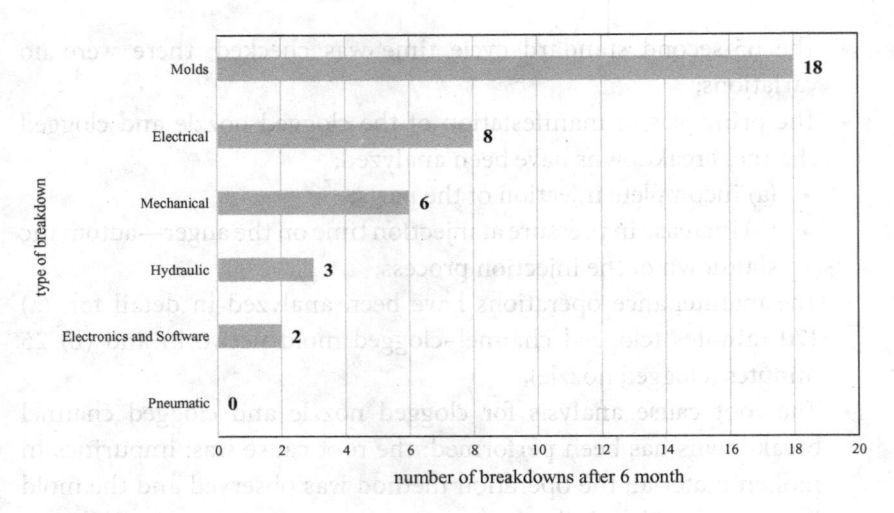

FIGURE 5.15
Results of improvements implementations for breakdowns after six months (molds: clogged nozzle and clogged channel).

- The MCPD project team has carried out the performance analysis of manufacturing improvement budget targets and the actual manufacturing budgets drawn up; the targets have been met;
- The MCPD project team carried out the evaluation of the PFC losses and waste performance indicators after implementing CCLW and CLW improvements (after 30 days, following MCI management process–see Section 3.4.3);
- The MCPD team leader of this project has conducted an evaluation of the degree of involvement of team members to achieve the MCI target of this project.

Step 7: Daily MCI Management: Standardization and Expansion of Validated Solutions

- A specific work instruction was performed to clean the injection nozzle;
- The team replicated the two improvements to six similar equipment.

Future actions:

- Improvements for set-up time have been planned to increase OEE levels for the horizontal plastic injection molding machine;

5.3.2 Improvement Project 6: Systematic MCI by Equipment Set-up, Settings, Adjustments Time Reduction

Company background:

- "FF-Plant" joined the MCPD system two years ago (Posteucă and Zapciu, 2013; Posteucă and Zapciu, 2015e);
- Type of industry: manufacturing and assembly; household appliances;
- Manufacturing regime: repeated lot–machine shop, plastic injection and assembly lines;
- PFC number: 4 (4 assembly lines);
- Improvement area: horizontal plastic injection molding machine ("AS").

5.3.2.1 Phase 1: Manufacturing Cost Policy Analysis

Step 1: From Company Productivity Vision to Establish Annual MCI Goal: Current State of the Equipment from the CCLW Perspective

- Annual MCI goal: $2,500,000;
- Number of MCI means (annual strategic improvement projects to meet annual MCI goal by CAMPT): seven *kaizen* projects and two *kaikaku* projects (similar to Figure 3.13);
- Stake to improve CLW associated with set-up, settings, adjustments, time improvement: 15% of annual MCI goal (third place as weight; one *kaizen* strategic project was selected out of seven).

Step 2: Target setting for CCLW: Full Understanding of the Current State of Losses and Waste

- After completing and carefully analyzing the 7 matrices (for PFC analyzed), annual CCLW related with set-up, settings, adjustments: $375,000 (including costs of WIP S; including stopping and slowing upstream and downstream processes of PFCs and system stops/slowdowns);
- The *kaizen* project theme is to reduce the average set-up, settings, adjustments losses on strategic equipment to under ten minutes; in this case it is the equipment "AS"—horizontal plastic injection molding machine:
  - The strategic interest of set-up: decrease set-up time to increase production and increase set-up number to increase flexibility;

- Process: injection of plastic parts–line 2;
- Cycle time: 45 seconds;
- Daily work schedule: three shifts, 1,315 minutes;
- Daily production (three shifts): 1,419 parts;
- OEE: 81% (see Figure 5.16);
- OEE target: 86% (after improving the set-up, settings, adjustments to meet the annual MCI target);
- Average set-up time: 32 min;
- Set-up target time: under 10 minutes (SMED);
- Set-up number per day: 4 events;

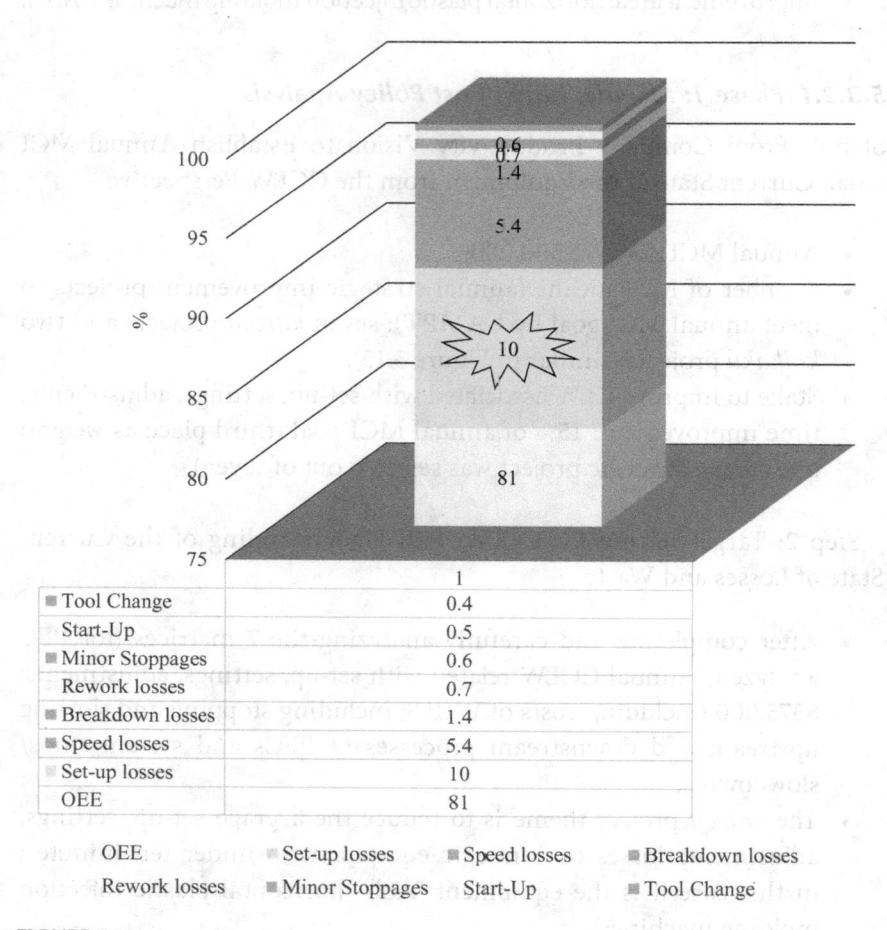

FIGURE 5.16

OEE structure for horizontal plastic injection molding machine (average measurements for the last six months).

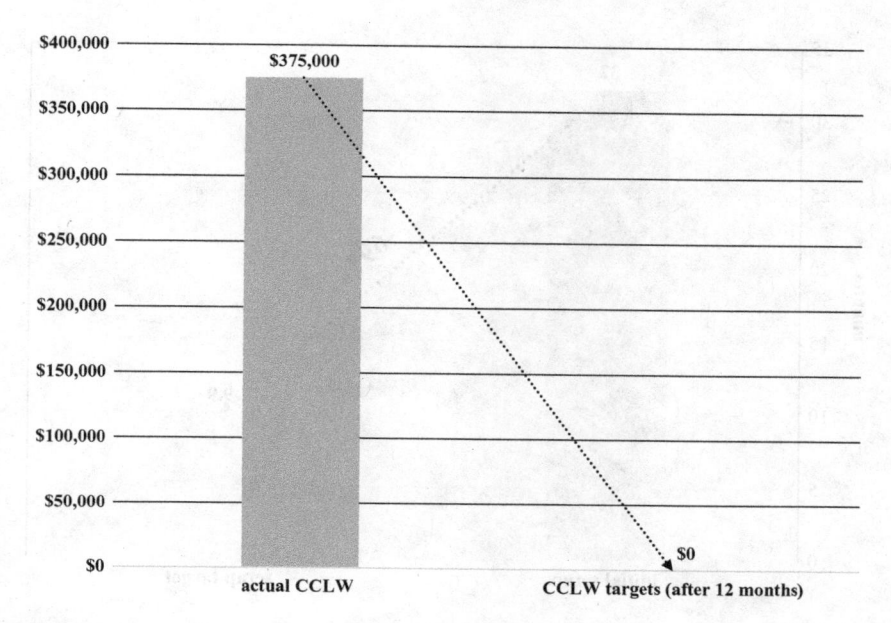

FIGURE 5.17
Setting annual CCLW target for set-up–horizontal plastic injection molding machine ("AS")

- Target number of set-up per day: 7 events (need for flexibility is required by production planning and customer requests);
- Missed production opportunity for each set-up event: 43 parts per set-up (the mold is for a part);
- Annual MCI targets for set-up, settings, adjustments losses for equipment "AS" (horizontal plastic injection molding machine) (see Figure 3.11 and Figure 3.12): improving costs of set-up losses; the need to reduce/ eliminate CCLW coordinates the need for improvement (see Figure 5.17);
- Annual MCI means target for set-up, settings, adjustments losses for equipment "AS" (horizontal plastic injection molding machine) (see Figure 5.18).

5.3.2.2 Phase 2: Manufacturing Cost Policy Development

Step 3: Annual Budgets for MCI: The Impact of CCLW Improvement Through this Project from the Annual MCI Goal Perspective

- The cost benefit analysis was performed in advance; the improvement costs were $4,500 for: (1) purchase of a mold preheating device; and

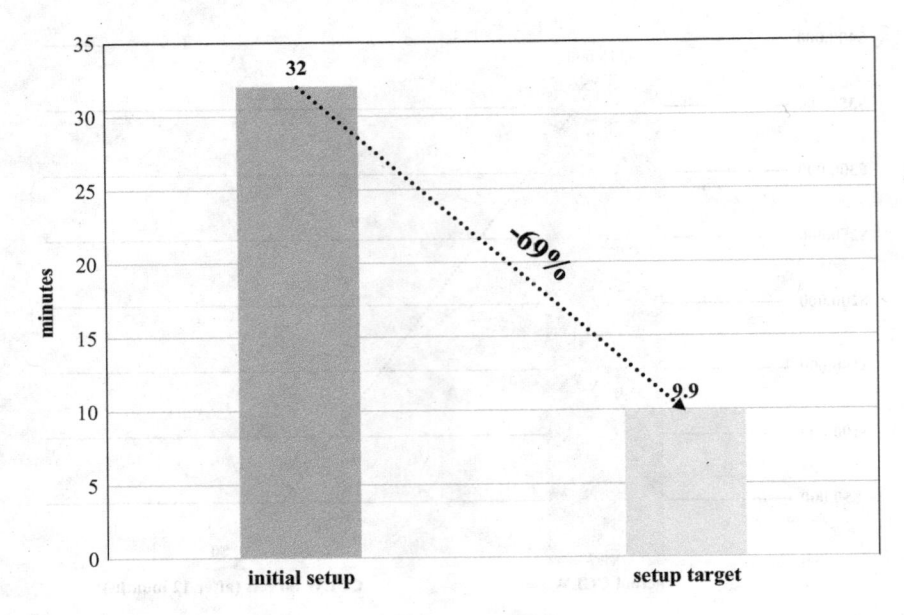

FIGURE 5.18

Setting annual MCI means target for set-up–horizontal plastic injection molding machine ("AS").

(2) mobile support for temporary storage of the old mold near the equipment; costs associated with *kaizen* team's hourly wage costs;

- The profitability of the missed opportunity to achieve production was assessed—the equivalent of parts for set-up: the average of 91 set-ups per month for "AS" represents 3,882 pieces of lost production opportunity (or 2.74 production days per month);
- The MCPD project team has carefully followed the manufacturing improvement budgets cycle (the five steps described in Section 3.3.1).

Step 4: Action Plan for MCI: From Setting the CCLW Target to Implementing the New Improvement Standard

- A team consisting of six members and a project leader was established;
- the two-week deadline was set to identify improvements, three weeks to implement improvements and another month to expand solutions to other similar equipment;
- The main tasks of the team were:
  - To perform an assessment of the current status of the "AS": horizontal plastic injection molding machine;

- To check the current status of the CLW associated to set-up, settings, adjustments losses for equipment "AS" and to check the current level of CCLW declared in the system;
- To prepare a plan of consistent measures for the elimination of costs of set-up, settings, adjustments losses;
- To propose countermeasures for issues identified at set-up, settings, adjustments time level;
- To define new standards of work set-up, settings, adjustments;
- To extend the time-validated solutions to other similar equipment.

5.3.2.3 Phase 3: Manufacturing Cost Policy Management

Step 5: Engage the Workforce for MCI: Analyzing and Identifying Improvement Solutions to Reach the CCLW Target

- The team analyzed in detail each parameter of "AS" and fully understood the principles of operation by phases;
- The team performed three sequential measurements for set-up time on "AS" (at different times of the day and with different setter teams; two setters plus one operator preparing the new raw material); average set-up time: 32 min (based on time study and video analysis);
- 3 major issues were identified:
  - 69% of set-up operations were performed by Setter 1 and 31% by Setter 2;
  - The waiting time for heating the new mold was 930 seconds (or 15.5 minutes);
  - The time for transferring the old mold to the storage rack and transferring the new mold to the 330 second (or 5.5 minutes)
- The spaghetti diagram for the two setters and the feeder for all set-up activities was done;
- Solutions chosen:
  - To reduce the heating time of the new mold, a boiler was chosen to pre-heat the mold outside the "AS" (the mold heating is done outside set-up time); thus the time of 15.5 minutes was eliminated;
  - To reduce the time spent on the transfer of the old mold to the storage area, a mobile support for the temporary storage of the old mold was purchased (near "AS"); the time of 5.5 minutes was reduced to one minute;

- The unnecessary movements/displacements of the setters and the feeder were reduced: the distance was decreased from 69 meters to 17 meters; the time was reduced by 3.3 minutes; the tasks for the two setters were balanced; the feeder has received some set-up tasks.

Step 6: MCI Performance: Checking the CCLW Improvements Results

- CCLW level for target setting, settings, adjustments for "AS" was met: $375,000;
- The *kaizen* project for MCI set-up fell within the initial deadline;
- The new standard for set-up settings, adjustments for "AS" is 9.2 minutes (all the time from the last compliant quality piece in the previous lot to the first quality piece in the next lot) (see Figure 5.19);
- Increased the opportunity to achieve production through saved time by reducing set-up time;
- The daily average of the "AS" set-up number increased from four to seven (increased flexibility for accepting small batches);
- The 86% OEE target was reached after improving the set-up, settings, adjustments to meet the annual MCI target;

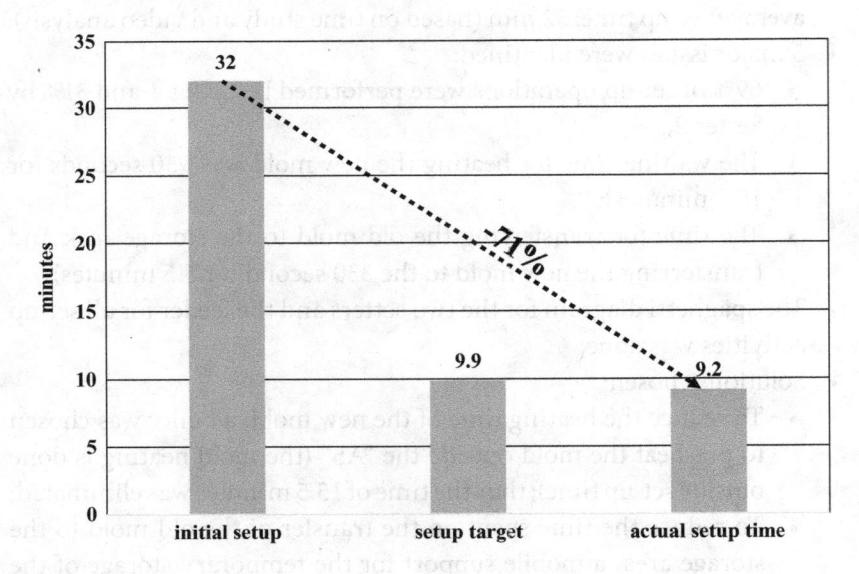

FIGURE 5.19
New set-up standard for "AS" equipment–overcoming the initial improvement target.

- The MCPD project team has carried out the performance analysis of manufacturing improvement budget targets and the actual manufacturing budgets drawn up; the targets have been met;
- The team has carried out the analysis of the performance of the cash flow improvement budget targets and the actual cash budgets drawn up; the targets have been met;
- The MCPD project team carried out the evaluation of the PFC losses and waste performance indicators after implementing CCLW and CLW improvements (after 30 days, following MCI management process –see Section 3.4.3);
- The MCPD team leader of this project has conducted an evaluation of the degree of involvement of team members to achieve the MCI target of this project.

Step 7: Daily MCI Management: Standardization and Expansion of Validated Solutions

- The team has developed and implemented a new SOP for set-up, settings, adjustments losses for the equipment "AS" (horizontal plastic injection molding machine);
- The MCPD team leader of this project has continuously updated the state of the MCI improvement project at the MCPD system information center.

Future activities:

- Enhancements have been planned for the other six similar pieces of equipment.

Six CCLW improvement projects were presented in this chapter to meet the annual MCI goal, given that companies were in need of maximizing manufacturing capacity (maximizing outputs "E") amid the increase in the number of products needed to be sold (see Figure 3.2). Improvements targeted SDI (1)–CLW 1 (see Matrices 1–7 from Figures 3.4, 3.9, 3.11 and 3.12).

6

MCI and Increasing Manufacturing Capacity through New Products and a New Equipment

The continuous improvement of company effectiveness to increase the use of current capabilities through the continuous development of new profitable products, and the further increase of equipment capacity through the purchase of new equipment represents a basic concern of MCI change drivers of the need to increase manufacturing capacity (maximizing outputs "E") to ensure external manufacturing target profit (see Figure 3.2).

Therefore, manufacturing companies need: (1) continuous knowledge of the LW levels of current equipment associated with the difference between value adding operating time (the time to make profitable products) and working hours, (2) CLW and (3) the continuous reduction of CCLW by planning and implementing annual MCI means at the required deadlines.

In this chapter, two systemic improvement projects (*kaikaku*) will be presented at "GG-Plant": a project to develop new profitable products that fit into the necessary time-to-market (TtM) and a project to install new equipment to fit the established time-to-start manufacturing (TtSM) of the new equipment. Improvements target SDI (1)–CLW 2 and CLW 3 (see Matrices 1–7 from Figures 3.4, 3.9, 3.11 and 3.12). Targeting of annual MCI targets for new products and new equipment is performed to increase the number of products manufactured and sold (see Figure 6.1).

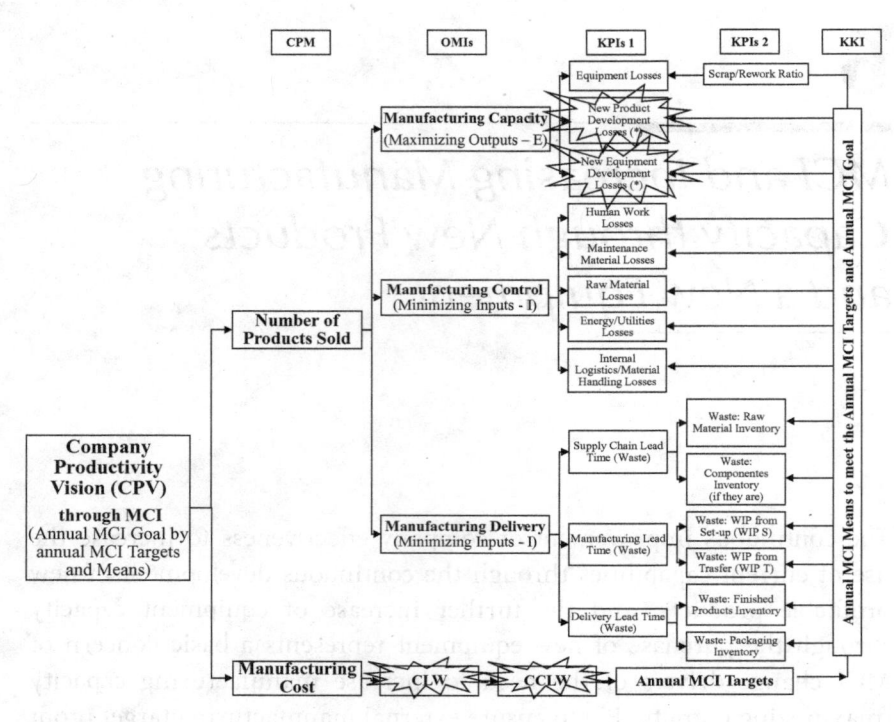

FIGURE 6.1
MCI policy deployment for a PFC for strategically targeting new products and new equipment development.

6.1 IMPROVEMENT PROJECT 7: SYSTEMIC MCI BY DEVELOPING A NEW IN-HOUSE PRODUCT

To achieve "doing the right things" (Drucker, 1963), among other things, the need for continuous and timely launch of profitable products (effectiveness) and then increased attention on efficiency to achieve "doing things right" (Drucker, 1963) is required. Unfortunately, when managers in many companies are concerned about productivity, they are almost exclusively focused on efficiency, or achieving "doing things right." The launch of new products and the installation of new equipment suitable for new products is the key to the strategic and systemic approach of CLW.

Any delay in TtM or the realization of a product with features not perceived as new are factors that cause CLW at the system level. In this context, unit manufacturing costs are higher than they could be for both the new product and the other products in that range and beyond.

Below is a brief description of an existing product improvement project to achieve a new product in the surgical medical devices industry for:

1. Ensuring a level of competitiveness imposed by the market (*doing the right things*) and
2. Ensuring an optimal charge on the company's current capabilities (*doing things right*).

- *Company background:*
  - "GG-Plant" joined the MCPD 3 years ago (Posteucă and Zapciu, 2015a; Posteucă and Zapciu, 2015b);
  - Type of industry: manufacturing and assembly; surgical medical devices;
  - Manufacturing regime: repeated lot–machine shop, plastic injection and assembly lines;
  - PFC number: 2 (two assembly lines);
  - The top management within the "GG-Plant" was aware that in order to sustain Maximizing Outputs ("E"), the development of new profitable products must meet two major needs:
    - Ensuring the level of competitiveness on the market: continuous and timely development of quality products with new features, with prices considered acceptable, through stimulation of a high volume of sales and low cost throughout the product life cycle (Akao, 1990);
    - Ensuring the optimal charging of current and future available capacities of the company (NCU): ensuring a synchronization between the volume of the products in the decline phase with the products in the marketing and growing phases (see Figure 3.10).

  Note: If one or both of these needs is not met, then the systemic appearance of CLW at process level is facilitated, both at TRL and PL (Posteucă and Sakamoto, 2017, pp. 173–174).
- At the same time, top management within "GG-Plant" understood that from the perspective of the MCPD system, the R & D process of new products takes into account the CLW level in similar processes of existing products to identify directions credible and capable of reducing costs with materials and workmanship, with a view to supporting a market segment indicated by the marketing department; more precisely the current state of MMIB. For example,

the level of CLW associated with equipment losses (beyond OEE) tends to remain stagnant at the level of the process or to increase if *kaizen* and/or *kaikaku* projects for MCI are not implemented (e.g. set-up time may remain unchanged for a piece of equipment regardless of what product being manufactured with that equipment; consequently CLW associated with set-up loss exists and can be anticipated, including at the CCLW level);

- Within "GG-Plant," the revolutionary products (with radically changed features), the ones that change the trend of the surgical medical devices industry, are products that are strategically approached from the perspective of development costs and unitary manufacturing costs and not only in terms of costs.

6.1.1 Phase 1: Manufacturing Cost Policy Analysis

Step 1: From Company Productivity Vision to Establish Annual MCI Goal

- Annual MCI goal: $8,500,000;
- Number of MCI means (annual strategic improvement projects to meet annual MCI goal by CAMPT): 12 *kaizen* projects and two *kaikaku* project (similar to Figure 3.13);
- Stake to improve CLW associated with new product: 18% of annual MCI goal (4th place as weight; one strategic *kaikaku* was selected out of two).

Step 2: Targets for the Development of the New Product and Target Setting for CCLW: Full Understanding of the Current State of Losses and Waste

- The process of research and development of new products started by conducting a competitive benchmarking study for: (1) prices and (2) characteristics of the new product (Posteucă, 2018, p. 60–74);
- The new "EAS" product must:
  - contribute to expanding the "DFD" product range;
  - continue the concept of current product;
  - be feasible;
  - be patented.

- The "EAS" target was to have a competitive price, the most competitive on a particular market. This could be accomplished through a high volume of outputs (maximizing outputs "E") and a continuous decrease in material costs and transformation costs (minimizing inputs "I"). In order to influence the level of price, which is often imposed by the market, the redesigning of existing products was the first step, taking into account all future opportunities to reduce CLW associated with the future manufacturing flow of "EAS" product. (Posteucă and Sakamoto, 2017, pp. 170–188; Posteucă, 2018, pp. 195–203);
- Since the future "EAS" product is not a completely new one, a revolutionary product, product research and development was started from the current status of KPIs related to existing products that are undergoing the redesign process (Ichida, 1996);
- The cost of investing in the new product was estimated at $3,850,000 (see Figure 6.2); most of the investment is in R&D for designing the new product (designing the new product, designing components, choosing materials and components to fit the target cost–the pressure on the cost tables to reach the target cost of materials for the new product);
- Following the analysis of the level of potential prices to be practiced over the life cycle of the product, analysis of potential sales volumes, current cost analysis with materials and with transformation costs, CLW analysis (MCI potentials throughout the product life cycle in terms of materials and, especially, transformation costs level), it was decided that the new "EAS" product should contribute to reducing/

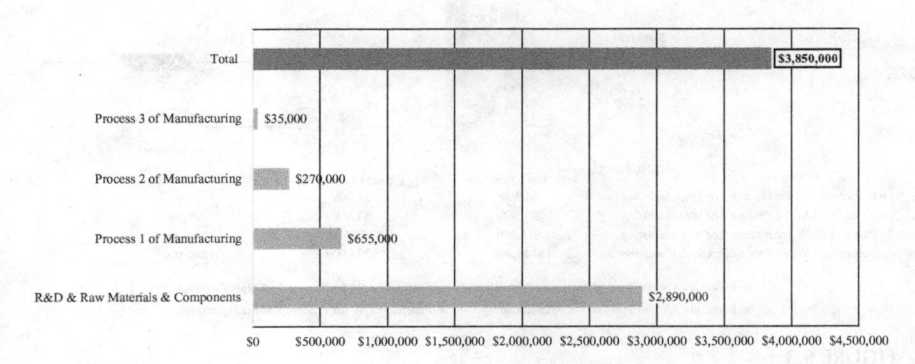

FIGURE 6.2
Structure of the investment budget by phases of the realization of the new product "EAS."

eliminating the current CLWs of existing products through an appropriate design;

- The possible CLWs to be reduced for the 30 months of the entire life cycle were estimated at the total amount of $450,000 (see Figure 6.3); this amount is the "hidden gold mine" from which the subsequent cost reductions will be exploited through the MCPD system to support the need for continued price reduction (especially in terms of declining sales volumes to meet the overall multiannual target profits level for the new "EAS" product–throughout its life cycle, eventually to help sustain the profitability of other products from the same PFC); this amount of $450,000 is included in the multiannual profit plan of the future "EAS" product, along with target sales revenue (external profit); just like the future costs of the new product can be forecasted, CLW can also be predicted. Some of the CLW-creating phenomena of current products, relatively similar to "EAS," will be preserved in the future;

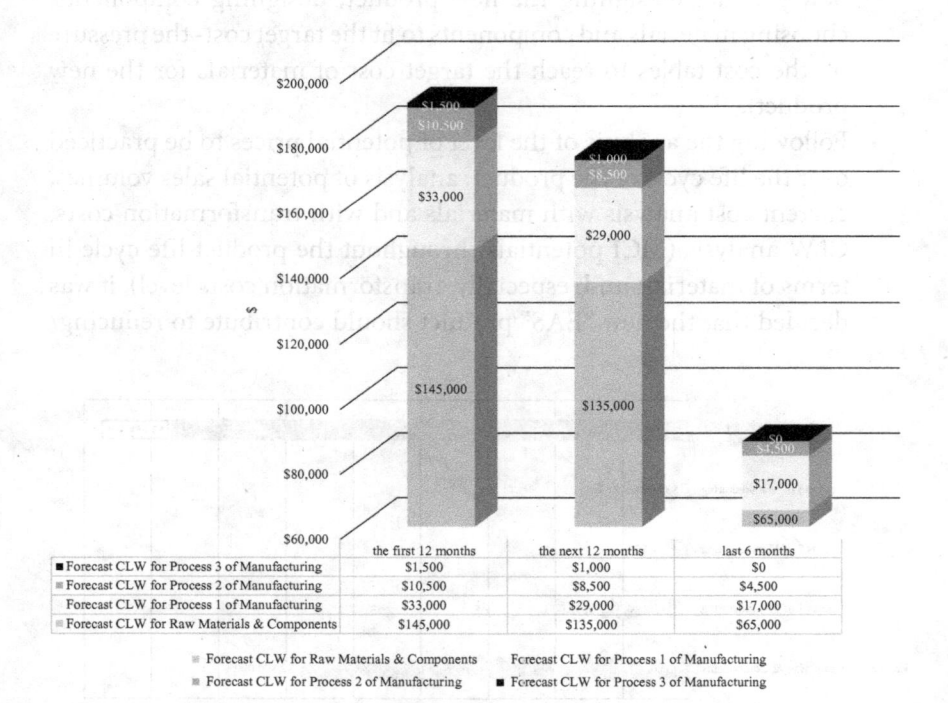

	the first 12 months	the next 12 months	last 6 months
■ Forecast CLW for Process 3 of Manufacturing	$1,500	$1,000	$0
■ Forecast CLW for Process 2 of Manufacturing	$10,500	$8,500	$4,500
Forecast CLW for Process 1 of Manufacturing	$33,000	$29,000	$17,000
■ Forecast CLW for Raw Materials & Components	$145,000	$135,000	$65,000

■ Forecast CLW for Raw Materials & Components Forecast CLW for Process 1 of Manufacturing
■ Forecast CLW for Process 2 of Manufacturing ■ Forecast CLW for Process 3 of Manufacturing

FIGURE 6.3
Estimation of possible CLW distribution across cost centers for the entire life cycle of the new product.

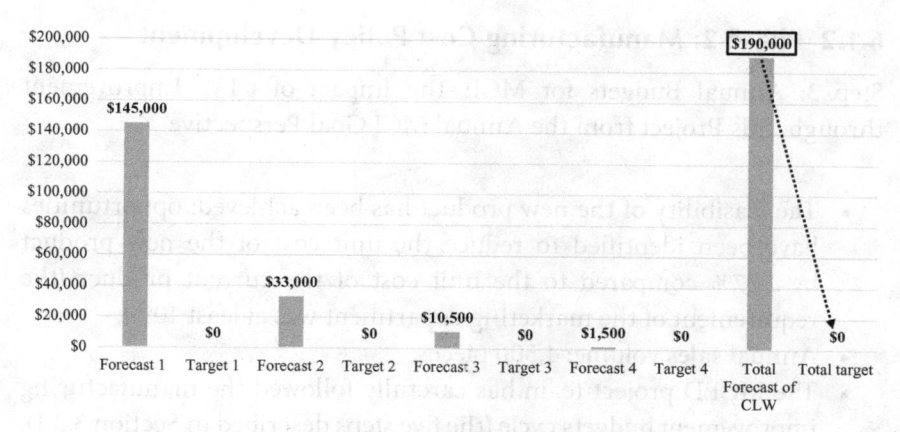

Legend:
Forcast 1: CLW for Raw Materials & Components
Target 1: Target for CLW of Raw Materials & Components
Forcast 2: CLW for Process 1 of Manufacturing
Target 2: CLW of Process 1 of Manufacturing
Forcast 3: CLW for Process 2 of Manufacturing
Target 3: CLW of Process 2 of Manufacturing
Forcast 4: CLW for Process 3 of Manufacturing
Target 4: CLW of Process 3 of Manufacturing

FIGURE 6.4
CLW targets distribution for each cost center (process) for the next 12 months from the launch of the new product ("EAS").

- For the next 12 months (as of March 1, the launch date of the new product), annual CLW related to new product development losses possible to eliminate was $190,000; this amount is made up of (see Figure 6.4):
  - (1) part of the CLW from the current product processes that will be replaced by the new "EAS" product;
  - (2) removing the current CLW through a design that is more effective and more efficient from the start.
- The management team and all the people involved in the launch of "EAS" were aware of the possibilities of exploiting the opportunity to remove CLW from the processes of the new product launched on time;
- As has been said, there are a number of MCI means at "GG-Plant" (annual strategic improvement projects to meet annual MCI goal by CAMPT): 12 *kaizen* projects and two *kaikaku* projects (similar to Figure 3.13); a *kaikaku* project for the introduction of "EAS" (this project); a *kaikaku* project for the installation of a new equipment to achieve the production for "EAS" (in process 1); and related to "EAS," five *kaizen* projects have been planned for MCI for PFC including "EAS."

6.1.2 Phase 2: Manufacturing Cost Policy Development

Step 3: Annual Budgets for MCI: The Impact of CLW Improvement through This Project from the Annual MCI Goal Perspective

- The feasibility of the new product has been achieved: opportunities have been identified to reduce the unit cost of the new product by 11.7% compared to the unit cost of the current product (the requirement of the marketing department was at least 10%);
- Annual sales volume: 4,500 pieces;
- The MCPD project team has carefully followed the manufacturing improvement budgets cycle (the five steps described in Section 3.3.1).

Step 4: Action Plan for MCI: From Setting the CLW Target to Implementing the New Improvement Standard

- A team of 10 members and a project leader (specialist in the new product development department) was established; team members were from the following departments: new product development; industrial engineering; designing devices in production; assembly line 1 and quality assurance;
- Time-to-market (TtM) has been set for a maximum of eight months;
- The main tasks of the team were:
  - To perform an assessment of the current status of products prior to the "EAS" product to identify their improvement opportunities (KPIs level related to losses and waste of previous products; see Table 3.3 and Table 3.4) (Posteucă and Sakamoto, 2017, pp. 119–122); at the same time, the archives of maintenance, quality and continuous improvement were analyzed;
  - To check the current status of the CLW associated with the products prior to the "EAS" product (see Table 3.5) and check the current level of the CCLW declared in the system for these earlier products (see Table
  - To prepare a plan of consistent measures to eliminate potential CLW and CCLW for "EAS"; especially for the next six months after the launch of the new product;
  - To propose immediate countermeasures for the identified problems at the potential CLW and CCLW for "EAS."

6.1.3 Phase 3: Manufacturing Cost Policy Management

Step 5: Engage the Workforce for MCI: Analyzing and Identifying Improvement Solutions to Reach the CLW and CCLW Targets for the Next 12 Months

- The team analyzed the unit costs of the new "EAS" product (materials costs and transformation costs, current CLW for previous products; successive simulations performed by the R&D department until the target cost has been reached);
- Costs of investment needed in manufacturing (new equipment, new processes, new software, new automation, new specializations for people, etc.) and in storage;
- Time-to-market target (TtM) (time planning 8 months ahead: (1) completing the new design of the new product; (2) implementation; (3) trial 1, 2 and 3; (4) serial production; and (5) sales;
- Feasibility study for new product;
- The volume of target sales throughout the product life cycle for each target market (sales department information; growth potential markets were analyzed; simulation of the price trend of competing products in the future was checked in order to understand potential pressure on the need to reduce unit costs for each EAS life cycle);
- The required productivity level (to ensure optimal use of current and future production capacity);
- The list of the new features of "EAS," design and materials and components were analyzed to achieve compatibility with all manufacturing flow processes for "EAS";
- Risks have been identified in each future "EAS" process (Design Failure Mode and Effect Analysis–DFMEA, Process Failure Mode Effects Analysis–PFMEA and Control Plan were performed);
- The team identified 11 major prototype issues (meeting the product design features, features imposed by the marketing department) and nine in processes (at process parameter level) and set the immediate action plan to reduce potential CLWs;
- CLW improvement teams have been deployed and implemented on time:
  - a *kaikaku* project for launching "EAS,"
  - a *kaikaku* project to install a new equipment to achieve the production for "EAS";

- The 5 *kaizen* projects for MCI for PFC of which "EAS" was part to reach the annual MCI target of $190,000.

Step 6: MCI Performance: Checking the CLW Improvements Results

- The CLW target for "EAS" was achieved for the first 12 months from launching: $190,000;
- New standards have been implemented as a result of *kaizen* projects;
- savings of 18.5% were achieved from the initial investment budget (i.e. $712,250 saving of the $3,850,000 planned)
- The number of "EAS" components has been reduced compared to the previous similar product (from 32 to 26);
- 76% of the 26 components of "EAS" are similar to those of the previous product;
- Total cycle time in "EAS" processes was reduced by 17% (using the 5 *kaizen* projects for MCI);
- Manufacturing lead time related to "EAS" was reduced by 12% in the first 12 months since launch; this reduction was enjoyed by all PFC products of which EAS is part (with the five *kaizen* projects for MCI);
- The "EAS" price was the lowest in the market compared to the competitors' product prices; it was on average 4.73% lower than the next product at the selling price level on the same sales market;
- Target volume targets in the first 12 months were exceeded by 11%;
- The performances achieved were obtained thanks to the new "EAS" design and meeting the CLW target of $190,000 in the first 12 months of launch; the total $450,000 CLW target was fully met in the first 24 months since the launch through the implementation of the solutions identified by the *kaizen* projects for MCI; *Kaizen* projects for MCI were not interrupted;
- The MCPD team leader of this project has conducted an evaluation of the degree of involvement of team members to achieve the MCI target of this project.

Step 7: Daily MCI Management: Standardization and Expansion of Validated Solutions

- The team has developed and implemented new SOPs for "EAS";
- The MCPD team leader of this project has continuously updated the state of the MCI improvement project at the MCPD system information center.

Future activities:

- New innovative design opportunities for future products have been identified.

Conclusion

Therefore, this MCI systemic improvement project, along with other MCI improvement projects, has met the annual MCI goal at "GG-Plant." "GG-Plant" has introduced a series of other systemic structural change projects such as: 5S expansion in all areas of the company (including office and warehouse areas), operative maintenance, preventive maintenance, quality maintenance, quality assurance, education and training, redesigning the cost and budgeting system. After releasing the new product (finishing the activities of the systemic improvement project for MCI), *kaizen* projects have begun to achieve MCI through systematic improvements of PFC processes. Some *kaizen* for MCI projects were already in progress and had immediate effects on the costs of the new product just released. "GG-Plant" has a symbiosis between the traditional budgeting and costing system and the AMIB (for existing and new products) and AMCIB.

6.2 IMPROVEMENT PROJECT 8: SYSTEMIC MCI BY INSTALLING NEW EQUIPMENT

As mentioned before, the launch of new products and the installation of new equipment suitable for new products and volumes of all products represents the key to the strategic and systemic approach of CLW. Any delay in TtSM (effectiveness) or the installation of a new equipment that does not meet the parameters required to achieve the necessary product characteristics or which is too expensive to exploit (efficiency) are the two main factors causing CLW at system level from the perspective of the new equipment. In this context, unit manufacturing costs are higher than could be for all products manufactured with that equipment (Posteucă, 2018, pp. 287–293; Posteucă and Sakamoto, 2017, pp. 214–224; Posteucă, 2015).

Below is a brief description of an installation design of new equipment by replacing an equipment considered obsolete in the surgical medical devices industry to: (1) ensure the level of effectiveness required by the market volumes (*doing the right things*); and (2) to ensure an optimal charge of the company's current capabilities (*doing things right*).

This project for installing new equipment is a continuation of the previous improvement project (project 7) to support the TtM of the "EAS" product. As presented in the previous project (project 7), $270,000 of the total investment of $3,850,000 have been allocated for investments in process 2 needed to achieve the new "EAS" product (see Figure 6.2). Of the $270,000, $230,000 was allocated to install a new equipment: the "SAE" machine. To support production for "EAS," the remaining $40,000 were planned to be invested in: software, automation, training for people, and storage devices in process area 2.

- *Company background:*
  - "GG-Plant" joined the MCPD 3 years ago;
  - Type of industry: manufacturing and assembly; surgical medical devices;
  - Manufacturing regime: repeated lot–machine shop, plastic injection and assembly lines;
  - PFC number: 2 (2 assembly lines);
  - The top management within "GG-Plant" was aware that to support maximizing outputs ("E") the timely installation of "SAE" must meet two major needs:
    - Ensuring the level of effectiveness imposed by market volumes and current and future competitiveness (Nakajima, 1988; Shirose, 1999; Suzuki, 1994);
    - Ensuring the optimal charging of future equipment capacities (NCU): ensuring a synchronization between the volumes of products in the decline phase with the products in the marketing and growing phases (see Figure 3.10).

Note: If one or both of these needs is not met, then the systemic appearance of CLW at process level is facilitated, both TRL and PL (Posteucă and Sakamoto, 2017, pp. 173–174).

6.2.1 Phase 1: Manufacturing Cost Policy Analysis

Step 1: From Company Productivity Vision to Establish Annual MCI Goal: Current State of the Equipment from the CCLW Perspective

- Annual MCI goal: $8,500.000;
- Number of MCI means (annual strategic improvement projects to meet annual MCI goal by CAMPT): 12 *kaizen* projects and two *kaikaku* projects (similar to Figure 3.13);
- Stake to improve CLW associated with new equipment: 10% of annual MCI goal (5th place as weight; one *kaikaku* strategic project was selected out of two).

Step 2: Targets for the New Equipment and Target Setting for CCLW: Full Understanding of the Current State of Losses and Waste

- The targets of the new "SAE" equipment are to include innovative technical solutions that meet the demand for the new "EAS" product and other "SAE" products with a low cost of labor, without breakdowns, without product defects and with a low maintenance cost over the lifetime of the product;
- The choice of the new "SAE" equipment has started with the planning of increasing demand for the new "EAS" product (Posteucă and Sakamoto, 2017, p. 217);
- The volumes of the new "EAS" product could be achieved only in a proportion of 69% with the old equipment in the first 12 months since the launch of "EAS" and the risk for non-quality was high; therefore, it was necessary to install new equipment before launching the new "EAS" product; the OEE level of the old equipment already reached 79%, and it was unfeasible to continue with *kaizen* projects for MCI to raise the current OEE level, especially as even an 85% OEE, considered as the maximum possible for the old equipment, could not provide the production volume for all products manufactured with that equipment and, especially, for the volume of "EAS";
- Top management decided to use the two pieces of equipment in parallel for three months to ensure the necessary production level; three months later, the production volumes of the old equipment were completely replaced by those made by "SAE";
- The initial level of the planned OEE of the "SAE" was 83%; the "SAE" capacity requirement for each shift was achieved for all six product types manufactured with it, including the new "EAS" product;
- The installation of the new "SAE" equipment must meet the following technical and investment targets:

- Cycle time: 125 seconds (compared to 181 seconds with the old equipment);
- OEE percentage: 83% (compared to 79% with the old equipment);
- Set-up time of 9 minutes (compared to 25 minutes with the old equipment);
- Man-hours per product of 0.28 (compared to 0.45 with the old equipment);
- 0% scrap rates (compared to 1.45% with the old equipment);
- Three-month time-to-start manufacturing (TtSM) (compared to five months with the old equipment);
- The annual CCLW target forecast for "SAE" was set at $180,000 (for the ten months of year "N"; the "SAE" installation completion date was March 1 year "N" to start "EAS" production) (see Figure 6.5);

Note: the cost level for one minute of equipment use is known and planned; losses and waste level for the equipment is planned from the KPIs history related to losses and waste and is adjusted by managers according to future business conditions (see Tables 3.3 and 3.4; Figure 3.4) (Posteucă and Sakamoto, 2017, pp. 119–122), the CLW level for the new equipment is based on the KPIs history for the CLW (see Table 3.5 and Figure 3.9) and the CCLW is predicted on the basis of historical managerial adjusted data for the future (see Figure 3.11);

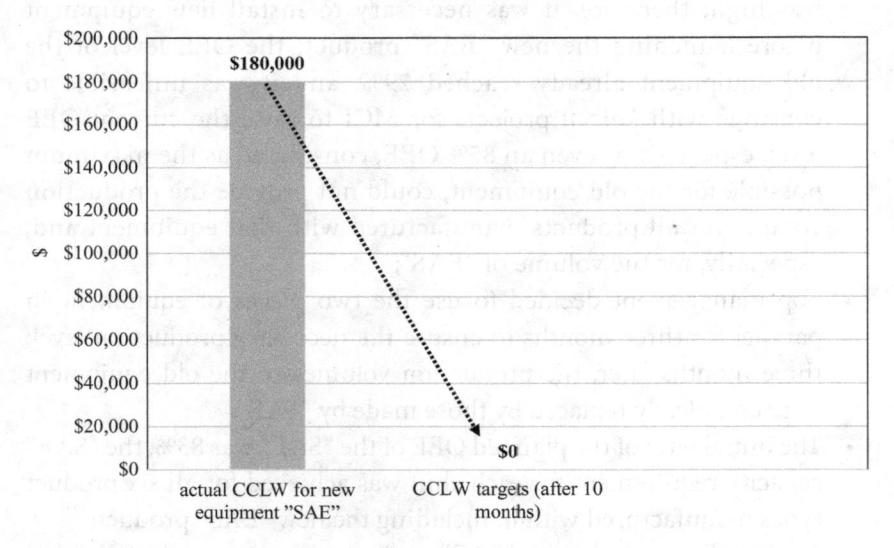

FIGURE 6.5

Setting annual CCLW target for the new equipment "SAE."

- Number of MCI means at "GG-Plant" (annual strategic improvement projects to meet annual MCI goal by CAMPT): 12 *kaizen* projects and two *kaikaku* projects (similar to Figure 3.13); a *kaikaku* project for the introduction of "EAS" (previous project; Project 7); a *kaikaku* project for installing new equipment to achieve production for "EAS" (in process 1; "SAE," this project); in connection to "EAS," five *kaizen* projects for MCI were planned for PFC of which "EAS" was a part (Project 7) and two *kaizen* projects for MCI to reduce "SAE" associated CLW (a *kaizen* for set-up, settings, adjustments time–establishing the SOP for the new equipment; a *kaizen* for loss of speed–establishing the SOP for the new equipment; following the two *kaizen* projects, the increase of the OEE for "SAE" is aimed from 83% to 84% in year "N" and reaching the annual CCLW target of $180,000).

6.2.2 Phase 2: Manufacturing Cost Policy Development

Step 3: Annual Budgets for MCI: The Impact of CLW Improvement Through This Project from the Annual MCI Goal Perspective

- The MCPD project team has carefully followed the manufacturing improvement budgets cycle (the five steps described in Section 3.3.1);
- The feasibility of the new equipment has been achieved: opportunities have been identified to reduce the unit cost of all products by 15.7% compared to the unit cost of the current product (the requirement of the marketing department was at least 10%);
- Annual sales volume for "EAS": 4,500 pieces;
- Three suppliers were identified for the "SAE" equipment;
- The *life cycle cost (LCC)* calculation for the new equipment was performed in a comparative manner for the three selected suppliers to simulate the subsequent unitary manufacturing cost.
  - The general information for all three versions of suppliers was the starting point:
    - Number of working days per year: 260 days;
    - A number of shifts per day: 3 shifts;
    - Years of operation of the equipment: 7 years;
    - The number of products to be produced per year: 158,000 pieces per year (including the 4,500 pieces of "EAS" from the first year; the trend of the new "EAS" is expected to increase more markedly in Year 2);

- The potential productivity analysis of each equipment provider was carried out:
  - Cycle time;
  - OEE percentage;
  - Set-up time;
  - Man-hours per product;
- An analysis of the potential quality of each equipment supplier was carried out:
  - Scrap and rework rate;
- The delivery time analysis has been performed for each equipment supplier:
  - Time-to-start manufacturing (TtSM);
- The initial costs were analyzed for each supplier:
  - Supplier cost;
  - Transportation cost;
  - Installation cost
  - Cost of testing;
  - The cost of replacing old equipment;
  - The decommissioning cost of "SAE"–at the end of the life cycle of the new equipment;
  - The residual value of the "SAE";
- The running costs for each supplier (the three selected suppliers) were analyzed:
  - Material costs/product;
  - Direct and indirect labor costs/product;
  - Utility costs/product;
  - Maintenance costs/product;
  - Spare parts costs/product;
  - Initial training costs/product;
  - Other costs.
- The supplier with the lowest unitary manufacturing cost and which best meets the technical and investment budget targets of "GG-Plant" has been selected.

Step 4: Action Plan for MCI: From Setting the CLW Target to Implementing the New Equipment

- A team consisting of ten members and one project leader was established;

- Time-to-start manufacturing (TtSM) has been set for a maximum of three months;
- The main tasks of the team were:
  - To check the current status of the CLW associated with current equipment that will be replaced by "SAE" (see Table) and check the current CCLW system level for this old equipment (see Table and the CLW and CCLW projections for "SAE";
  - To conceive a plan of consistent measures to eliminate potential CLW and CCLW for "SAE," especially for the next six months after installing the new equipment and launching the new "EAS" product;
  - To propose immediate countermeasures for the problems identified at the level of CLW and CCLW potential for "SAE" (hence the *kaizen* project for set-up, settings, adjustment time—establishing SOP for the new equipment—and the *kaizen* project for loss of speed; the two *kaizen* projects aimed at increasing the OES for the "SAE" from 83% to 84% in the "N" year and reaching the annual CCLW target of $180,000).

6.2.3 Phase 3: Manufacturing Cost Policy Management

Step 5: Engage the Workforce for MCI: Analyzing and Identifying Improvement Solutions to Reach the CLW and CCLW Targets for the Next 12 Months;

- *Planning concept–15 days:*
  - A team analyzed the unit costs of the new "EAS" product and other products made with "SAE" based on the LCC analysis;
  - Time-to-start manufacturing (TtSM) of maximum three months: (1) choosing the supplier for the new equipment; (2) transport; (3) installation; (3) testing of new equipment; and (4) serial production;
  - The team members of the *manufacturing engineering department* and the *maintenance department* have defined the basic characteristics of the "SAE": equipment size; equipment supply; part discharge system; parameters and operating principles of the "SAE"; the typology of consumables and their frequency of replacement; autonomous maintenance standards (equipment points and frequency for cleaning, inspection and lubrication);

the schedule for planned maintenance (replacement parts management based on the initial and subsequent frequency of change of parts); hourly productivity levels; identifying potential points of the new equipment mechanism that can lead to scrap and rework; types of training required by operators to operate the new equipment and maintain it; utilities consumption, etc. Team members of the *manufacturing engineering department* and the *maintenance department* have defined the following for the "SAE": reliability; maintainability; operability; quality; flexibility; and energy consumption.

- *Action plan and design–15 days:*
  - The team chose the supplier for "SAE"–the "AAA" supplier; the choice was approved by the top management within "GG-Plant";
- *Fabrication–30 days:*
  - Executing the equipment and testing it by the "AAA" supplier, in accordance with all the requirements of the "GG-Plant," lasted for 30 days. A team of three specialists from the company "GG-Plant" made two visits to the "AAA" factory to check the design of the equipment and how to meet the 3 "SAE" specific requirements—additional requirements to the standard features of equipment currently manufactured by "AAA";
- *Transport and installation of new equipment–seven days:*
  - The "SAR" transport lasted for four days. Installation was carried out by specialists from "AAA" and lasted for three days. Prior to installation, layout and dismantling of the old equipment were performed. The dismantling of the old equipment was performed by employees of the company "GG-Plant."
- *Testing of products and start of production–ten days:*
  - Product testing and operator training lasted ten days.

Step 6: MCI Performance: Checking the CLW Improvements Results

- The team deployed and implemented the "SAE" installation *kaikaku* project on time: time-to-start manufacturing (TtSM) was 77 days (2.5 months compared to the initial target of three months);
- The *kaizen* teams have deployed and implemented in time the two *kaizen* projects for MCI in the ten months of year "N" (a *kaizen* for set-up, settings, adjustments time—establishing SOP for the new equipment—and a *kaizen* for loss of speed;

- Following the two *kaizen* projects, the increase of the OEE for the "SAE" was reached from 83% to 84% and an annual CCLW target of $180,000;
- New standards resulting following the *kaizen* projects (SOP for set-up and cycle time) have been implemented;
- Savings from the initial investment budget of 25% (i.e. a saving of $57,500 out of the $230,000 planned to install a new "SAE") were achieved);
- Target volume targets in the first 12 months were exceeded by 11%;
- The MCPD team leader of this project has conducted an evaluation of the degree of involvement of team members to achieve the MCI target of this project.

Step 7: Daily MCI Management: Standardization and Expansion of Validated Solutions

- The team has developed and implemented new SOPs for "SAE" related to set-up and speed.

Future activities:

- The *kaizen* projects for MCI affiliated to "SAE" have not been discontinued;
- The MCPD team leader of this project has continuously updated the state of the MCI improvement project at the MCPD system information center.

Therefore, two CCLW systemic improvement projects were presented in this chapter in order to meet the annual MCI goal, given that "GG-plant" was in need of manufacturing capacity (maximizing outputs–"E") amid the increase of the number of products to be sold (see Figure 3.2) by planning and implementing a new product and new equipment. Improvements targeted SDI (1)–CLW 2 and CLW 3 (see Matrices 1–7 from Figures 3.4, 3.9, 3.11 and 3.12).

Part III

Profitable Improvement Projects Implementation through MCPD for MCI toward Minimizing Inputs

This third part of this book addresses the implementation of profitable CCLW improvement projects through MCPD to meet MCI in order to consistently support internal manufacturing profit through minimizing inputs (hypostasis of "I" scenario; see Table 2.1).

This third part of the book will present systematic (*kaizen*) and systemic (*kaikaku*) improvement projects, or in other words, MCI means to meet MCI targets on time, based on credible and capable assumptions for each MCI strategy for change drivers (CD) of MCI to ensure internal manufacturing profit:

- CD4—Maximizing variable cost efficiency;
- CD5—Maximizing fixed cost efficiency;
- CD6—Continuously improving manufacturing lead time;

- CD7—Continuously aligning processes to market needs;
- CD8—Continually improving inventory levels.

The first chapter of the second part of this book, Chapter 7, presents actual systemic (*kaikaku*) improvement projects or annual MCI means (SDI 2–CLW 4–8; see Matrix 7 of Figure 3.14) for achieving the annual MCI targets (SDI 2–CLW 4–8; see Matrix 6 from Figure 3.12) converging to annual MCI goal (and CAMPT) by improving or eliminating CCLW to achieve *improving variable and fixed costs (material, environment and labor costs)*. Therefore, Section 7.1 presents the approach method and the results of a *kaikaku* project for MCI (systemic improvement) by replacing raw material for an existing product to reduce variable costs with materials. Section 7.2 describes the approach method and the results of a *kaikaku* project for MCI (systemic improvement) by reducing environmental costs to reduce relative fixed costs with the environment. Section 7.3 describes the approach method and results of a *kaikaku* project for MCI (systemic improvement) by implementing the simple automation device to reduce fixed costs with direct labor.

The second chapter of this second part of the book, Chapter 8, the last chapter of this book, presents four *kaizen* projects for real MCI or annual MCI means (SDI 3–CLW 9–14; see Matrix 7 of Figure 3.14) to achieve the annual MCI targets (SDI 3–CLW 9–14; see Matrix 6 from Figure 3.12) converging to the annual MCI goal (and CAMPT) to achieve *improving manufacturing lead time and aligning the manufacturing process to takt time*. Therefore, Section 8.1 presents the approach method and the results of a *kaizen* project for MCI (systematic improvement) by reducing the lead time of an assembly line. Section 8.2 presents the approach method and the results of a *kaizen* project for MCI (systematic improvement) by synchronizing the cycle time of the equipment with the assembly line cycle time. Section 8.3 presents the approach method and results of a *kaizen* project for MCI (systematic improvement) by reducing inventories of raw materials. Finally, Section 8.4 presents the approach method and the results of a *kaizen* project for MCI (systematic improvement) by reducing inventory of finished products.

7

MCI through Improving Variable and Fixed Costs: Material, Environment and Labor Costs

Continuous improvement of company efficiency through continuous improvement of variable and fixed costs is a basic concern of MCI change drivers to achieve *manufacturing control* (see Figure 3.2). Therefore, once a manufacturing company has made sure it has developed the most profitable products (CD2), that it has installed the most suitable equipment (CD3) and that this equipment has an appropriate level of effectiveness at the level of planned production (CD1), it will need to ensure strict control over all the resources consumed throughout the manufacturing flow of each PFC level (more precisely, at the level of variable costs—especially for indirect costs and fixed costs, and especially at direct labor costs, if they are handled as fixed).

Continuous knowledge and improvement of the LW, CLW and CCLW levels associated with the lack of efficient use of all resource inputs (variable cost and fixed cost) requires the planning and implementation of the annual MCI means (*kaizen* and *kaikaku* projects) at the necessary deadlines. Directing the annual MCI targets to increase the variable and fixed costs is carried out to increase the number of products manufactured and sold based on a continuous low cost (see Figure 7.1), especially amid sales volume reductions.

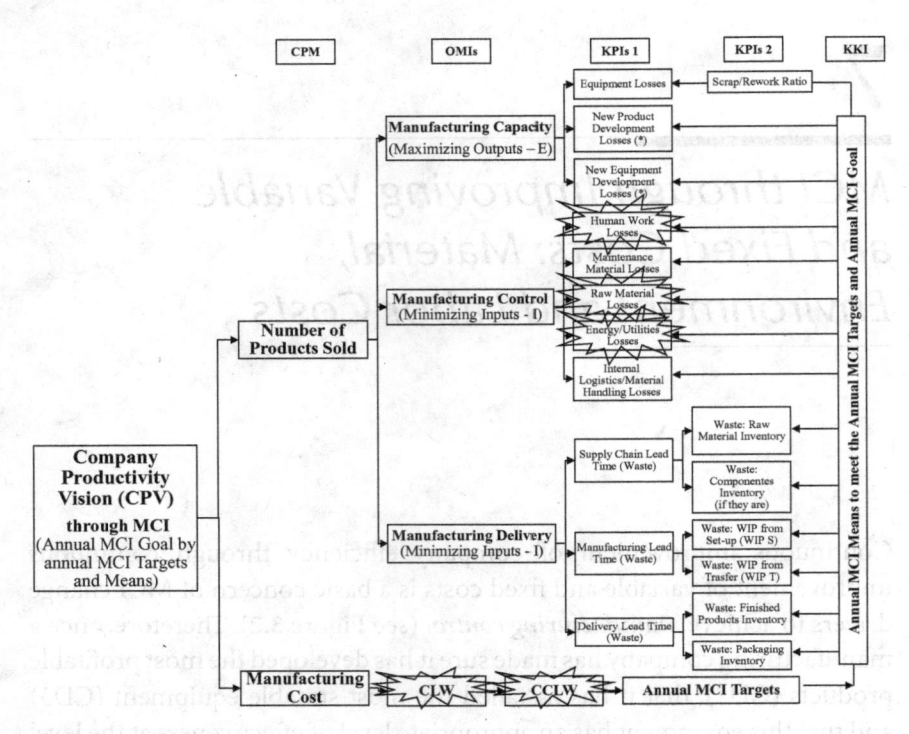

FIGURE 7.1
MCI policy deployment for a PFC for manufacturing control.

7.1 IMPROVEMENT PROJECT 9: SYSTEMIC MCI BY REDUCING A RAW MATERIAL, UTILITY AND LABOR COSTS FOR AN EXISTING PRODUCT

Company background:

- "HH-Plant" joined the MCPD system four years ago;
- Type of industry: manufacturing and assembly–electronic home appliances;
- Manufacturing regime: repeated lot–machine shop and assembly lines;
- PFC number: 4.

Defining the problem:

- Nine months after the launch of the "AAP" product in product family "B," the product price level became uncompetitive;

- "HH-Plant" top management was surprised by this market situation because the product life cycle of the "AAP" product was 24 months and they did not have at that time any new product to replace "AAP" to support the annual and multiannual profit plan;
- Based on the headquarters' marketing department's requests to reduce unit costs of "AAP" product by a minimum of 10% within a maximum of 3 months, "HH-Plant" top management has decided to re-analyze all unit costs of "AAP" product in order to identify consistent solutions to reduce unit costs by at least 12% for raw materials and components and at least 9% for utilities and direct labor;
- Improvement areas: R&D, procurement, manufacturing engineering, quality and maintenance.

7.1.1 Phase 1: Manufacturing Cost Policy Analysis

Step 1: From Company Productivity Vision to Establish Annual MCI Goal: Current State of the Product "AAP" from the CCLW Perspective:

- Annual company MCI goal: $5,500,000;
- Number of MCI means (annual strategic improvement projects to meet annual MCI goal by CAMPT): 11 *kaizen* projects and 2 *kaikaku* project (similar to Figure 3.13);
- Annual cost reduction target for "AAP": $100,000 per year.

Step 2: Target Setting for CCLW: Full Understanding of the Current State of Losses and Waste of Raw Material, Utility and Labor Costs Reduction of Product Family

- The theme of the *kaikaku* project for raw material, utility and labor costs reduction for the product "AAP": identifying the most feasible ways to reduce unit costs for "AAP";
- The CCLW annual product family of the "AAP" product family cannot provide a unit cost reduction of 10% or $100,000 per year for the total annual sales volume planned; the current level of CCLW is only $45,500 per year for all CLWs related to "AAP";
- The annual CCLW target planned for the current year for the "AAP" product was $11,500 (a part of the total product family of the

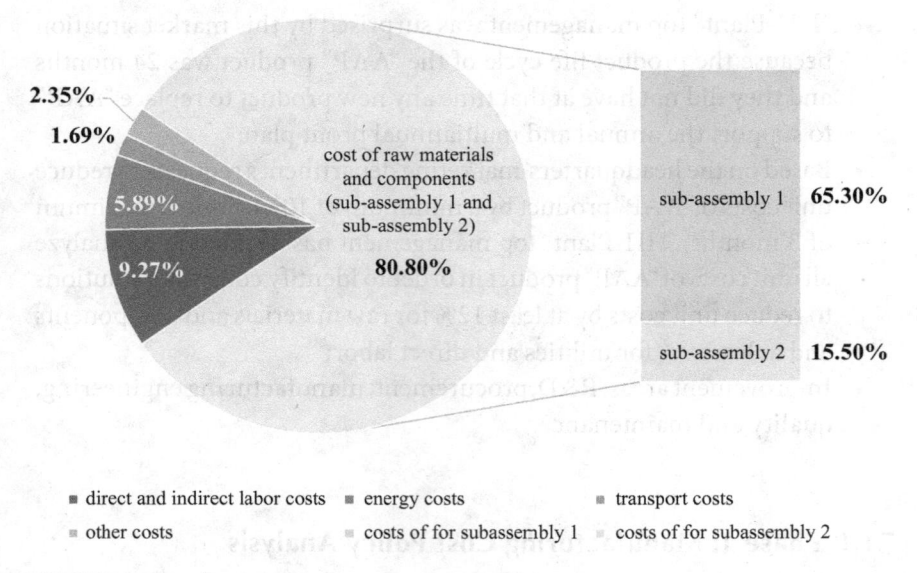

FIGURE 7.2
Percentage structure of unit costs of the product "AAP."

"AAP" product; one *kaizen* project for equipment losses–reducing breakdown time on "Y" equipment in Process 4);

- Information on the percentage structure of "AAP" product costs (see Figure 7.2):
- Annual MCI targets for "AAP" products to achieve a 10% unit cost reduction of $100,000 per year:
  - Annual CCLW target (1): $11,500 (1 *kaizen* project already planned at the beginning of the year for equipment losses–reducing breakdown time on equipment "Y" in Process 4);
  - Annual CCLW target (2): $30,000 (one new *kaizen* project for equipment losses–performance loss or cycle time reduction for equipment "Z" of Process 3 to increase productivity level for the "AAA" product and reduce unit cost; equipment "Z" was a bottleneck;
  - Redesign of the "AAP" product (replacing some of the raw materials and components currently used with others that are less expensive but perceived by customers to be of the same quality level): $47,500;
  - Cost reduction with utilities: $1,750;
  - Reducing direct and indirect labor costs: $9,250.

7.1.2 Phase 2: Manufacturing Cost Policy Development

Step 3: Annual Budgets for MCI: The Impact of CCLW Improvement and Cost Reduction from Redesign and Utilities and Labor through This Project from the Annual MCI Goal Perspective

- The cost-benefit analysis was performed in advance; the improvement costs were mainly associated with the hourly cost of the four *kaizen* teams specifically created to reduce the unit costs of the "AAA" product;
- The profitability of potential output volume losses for the "AAP" product was analyzed;
- "HH-Plant" top management decided that the four *kaizen* projects were strategic and of maximum importance;
- The MCPD project team has carefully followed the manufacturing improvement budgets cycle (the five steps described in Section 3.3.1).

Step 4: Action Plan for MCI: From Setting the CCLW Target to Implementing the New Improvement Standard

- Four *kaizen* teams were established;
  - Team 1: *kaizen* project for equipment losses–reducing breakdown time on equipment "Y" in Process 4 (five members and one project leader);
  - Team 2: new *kaizen* project for equipment losses–performance losses or cycle time reduction for equipment "Z" in Process 3 (seven members and one project leader);
  - Team 3: redesign of the AAP product (six members and one project leader);
  - Team 4: cost reduction for utilities and direct and indirect labor costs (eight members and one project leader).
- The 45-day deadline has been set to identify possible improvements for all four *kaizen* projects and to fully implement them within a maximum of 35 days.

7.1.3 Phase 3: Manufacturing Cost Policy Management

Step 5: Engage the Workforce for MCI: Analyzing and Identifying Improvement Solutions to Reach the CCLW and Cost Reduction Targets

- Team 1: *kaizen* project for equipment losses–reducing breakdown time in Process 4:

- The standard and current parameters of equipment "Y" have been checked;
- The principles of breakdown time were analyzed;
- The maintenance operations for breakdown of equipment "Y" have been analyzed in detail;
- An analysis of root causes for breakdown of equipment "Y" has been carried out;
- Three necessary improvements resulting from breakdown analysis of equipment "Y" have been identified;
- Team 2: new *kaizen* project for equipment losses–performance losses or cycle time reduction for equipment "Z" in Process 3:
  - The parameters of equipment "Z" were analyzed;
  - The time sequence that was a bottleneck within the equipment cycle time was identified;
  - The bottleneck cycle time for the time sequence was analyzed to identify improvement opportunities;
  - Two opportunities to reduce the equipment cycle time and to include cycle time in the new predetermined takt time have been identified.
- Team 3: redesign of the "AAP" product:
  - The current design of the "AAP" product was analyzed in detail:
    - Subassembly 1;
    - Subassembly 2;
  - Subassembly 1 was chosen to be analyzed in detail to identify substitution opportunities for raw material and components;
  - The following detailed information was identified for the subassembly 1:
    - Materials and components: 67 types of materials and 17 components;
    - Average purchase prices of the 67 types of materials and 17 components in the last 9 months;
    - Comparison of purchase prices and actual volumes with the initial standard prices and volumes; price increases have been identified for ferrous materials and some electronic components;
    - All 64 suppliers of materials and components and their contractual conditions were analyzed;
    - The cost of transport and intermediate storage was analyzed;
    - The current technical performances of subassembly 1 were analyzed;

- Two new operating principles of subassembly 1 (reverse engineering for the two products of the direct competitors) have been identified;
- Three new materials that could be used for subassembly 1 have been identified;
- The design for subassembly 1 has been remade;
- Two new suppliers were identified for the three new materials and the framework agreements were signed under favorable conditions for "HH-Plant."
- Team 4: cost reduction for utilities and direct and indirect labor costs:
  - Process time (cycle time and process lead time) was analyzed for all 14 main processes related to the manufacturing and assembling flow of "AAP" product;
  - Opportunities to reduce utility costs have been identified in processes 4, 8, 10 and 14;
  - Opportunities to reduce transport times and associated costs have been identified in processes 1, 5, 8, 10, 12 and 14 (reduce the transport time by 630 seconds for each "AAP" product);
  - The total annual cost of direct and indirect labor of the "AAP" product was analyzed for each process;
  - The time and motion study has been redone, and a number of cycle time improvement opportunities have been identified;
  - Method Design Concept (MDC) was used for the "AAP" product assembly line to directly reduce labor costs through a new design of work items; the results of the MDC project revealed a 27% reduction in direct labor costs.

Step 6: MCI Performance: Checking the CCLW Improvements Results

- Annual CCLW target (1) of $11,500 of *kaizen* project for equipment losses (reducing breakdown time on equipment "Y" in Process 4) has been achieved;
- Annual CCLW target (2) of $30,000 of the new *kaizen* project for equipment losses (performance loss or cycle time reduction for the equipment "Z" in Process 3) has been achieved;
- Reduction target of $47,500 through redesign of AAP product has been exceeded; the actual reduction was $65,700;

- Utilities cost reduction target of $1,750 has been exceeded; the actual reduction was $2,130;
- Cost reduction target with direct and indirect labor costs of $13,250 has been achieved (by reducing transport times, but especially through the MDC project–reducing the auxiliary function by setting the improvement target for reducing the auxiliary function consistently to the need for the annual reduction of $13,250–*kaizenshiro*) (Posteucă and Sakamoto, 2017, pp. 328–331);
- The total annual reduction achieved for "AAP" product was $122,580 (see Figure 7.3);
- The MCPD project team has carried out the performance analysis of manufacturing improvement budgets targets and the actual manufacturing budgets drawn up; the targets have been met;
- The team has carried out the analysis of the performance of the cash flow improvement budget targets and the actual cash budgets drawn up; the targets have been met;
- The MCPD project team carried out the evaluation of the PFC losses and waste performance indicators after implementing CCLW and CLW improvements (after 30 days, following MCI management process–see Section 3.4.3);
- The MCPD team leader of this project has conducted an evaluation of the degree of involvement of team members to achieve the MCI target of this project;

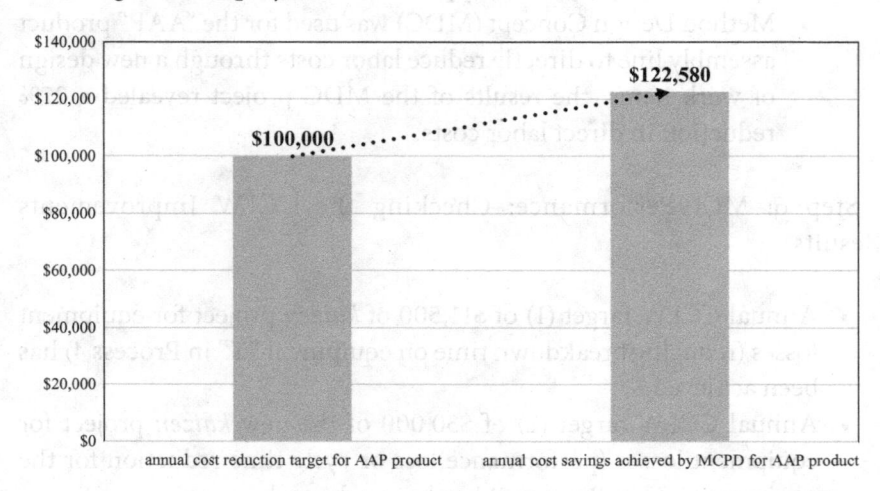

FIGURE 7.3

Annual cost reduction target versus annual cost savings achieved through MCPD for AAP product.

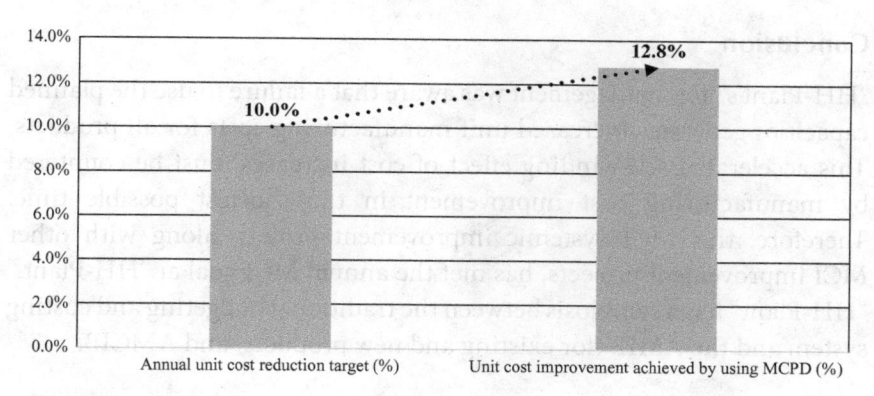

FIGURE 7.4

Percentage annual cost reduction target versus percentage reduction in unit cost obtained by using MCPD for "AAP" product.

- The percentage reduction target of 10% of the manufacturing and assembling unit cost has been achieved and exceeded. The annual reduction was 12.8% (see Figure 7.4).

Step 7: Daily MCI Management: Standardization and Expansion of Validated Solutions

- The management team has approved all the improvements proposed by the four *kaizen* teams;
- Teams 1 and 2 have developed and implemented the new standards for equipment breakdown and cycle time;
- Team 3 changed the design of the "AAP" product;
- Team 4 has implemented the new standards for utilities (energy) and labor costs;
- The MCPD team leader of this project has continuously updated the state of the MCI improvement project at the MCPD system information center.

Future activities:

- The *kaizen* teams have also identified new cost reductions opportunities for the "AAP" product, apart from those already approved by top management for the current 10% reduction requested by the marketing;
- Extensions of design improvement and cycle time from the assembly line to three other products similar to the "AAP" product have been planned.

Conclusion

"HH-Plant's" top management was aware that a failure to use the planned capacities generates increased unit manufacturing costs for all products. This accelerated self-winding effect of cost increases must be countered by manufacturing cost improvement in the shortest possible time. Therefore, this MCI systemic improvement project, along with other MCI improvement projects, has met the annual MCI goal at "HH-Plant." "HH-Plant" has a symbiosis between the traditional budgeting and costing system and the AMIB (for existing and new products) and AMCIB.

7.2 IMPROVEMENT PROJECT 10: SYSTEMIC MCI BY REDUCING ENVIRONMENT COSTS

Company background:

- "JJ-Plant" joined the MCPD system 18 months ago;
- Type of industry: process–ferrous and non-ferrous metal processing;
- Manufacturing regime: continuous;
- PFC number: 1;
- Improvement area: machine shop.

Defining strategic issues to support the long-term profit plan:

- Top management at "JJ-Plant" had four major concerns (Posteucă, 2013):
- Extending as much as possible the number of days of the continuous manufacturing flow;
- Improving OEE (especially shutdown and downtime);
- Reducing the electricity consumption;
- Reducing environment costs.

7.2.1 Phase 1: Manufacturing Cost Policy Analysis

Step 1: From Company Productivity Vision to Establish Annual MCI Goal: Current State of the Process/Equipment from the CCLW Perspective

- Annual MCI goal: $3,750,000;

- Number of MCI means: 8 *kaizen* projects and 1 *kaikaku* project–this project (similar to Figure 3.13);
- Stake to improve CLW associated with reducing environment costs: 18% of annual MCI goal.

Step 2: Target setting for CCLW: Full Understanding of the Current State of Losses and Waste

- After completing and carefully analyzing the 7 matrices (for PFC analyzed), annual CCLW related with reducing environment costs (chemical compound "X"): $675,000;
- The theme of the *kaikaku* project for reducing environment costs: reducing environmental costs by recycling the chemical compound "X";
- Information on "JJ-Plant" manufacturing process:
  - Number of distinct processes: 5 processes;
  - Number of processes in which the chemical compound "X" is used: 1 process (Process 5)
  - Evolution of consumption of chemical compound "X" in the last 7 months of year "N-1" (see Figure 7.5);

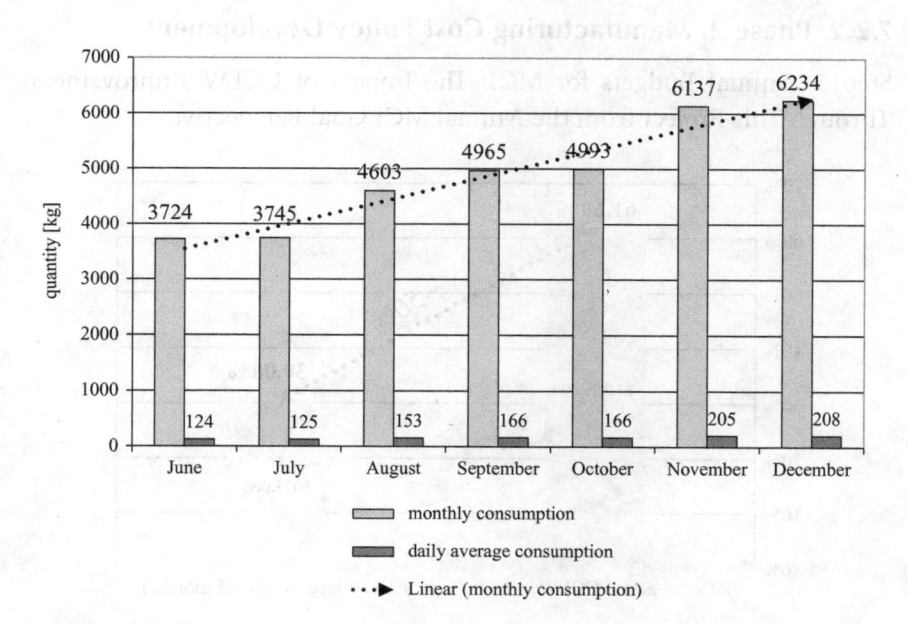

FIGURE 7.5

Quantitative evolution of chemical compound "X" consumption in the last seven months of year "N-1."

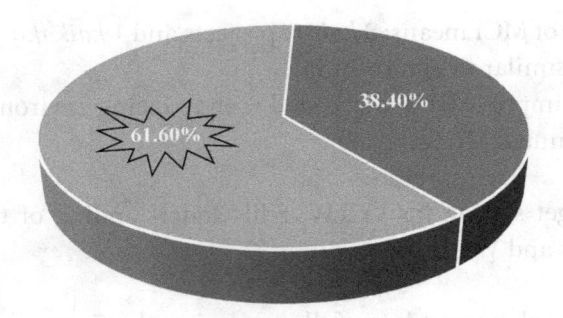

Residues of chemical compound "X" recovered by decantation
Residues of non-recoverable chemical compound "X" (external recycling)

FIGURE 7.6
Residues of the chemical compound "X."

- Chemical "X" compound residues for the last 7 months are shown in Figure 7.6.
- Annual MCI targets for reducing environmental costs by recycling the chemical compound "X" (see Figure 7.7); the amount of $1,350,000 is the cost of external recycling of chemical compound "X" (including: transportation, intermediate storage, handling, taxes, etc.).

7.2.2 Phase 2: Manufacturing Cost Policy Development

Step 3: Annual Budgets for MCI: The Impact of CCLW Improvement Through This Project from the Annual MCI Goal Perspective

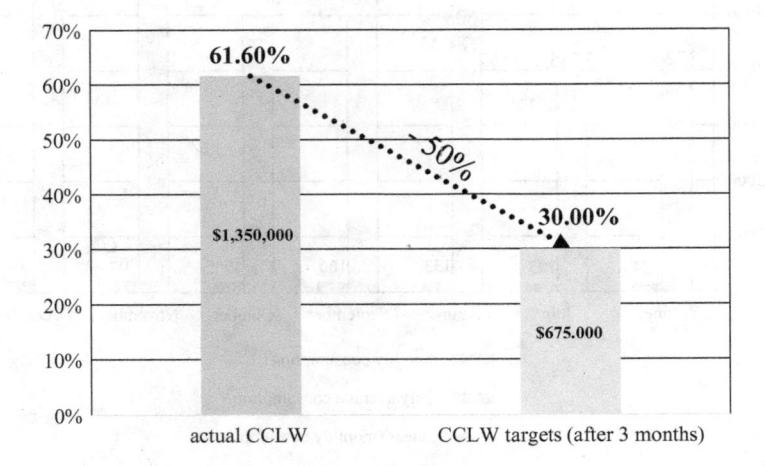

FIGURE 7.7
Setting annual CCLW target for reducing environmental costs by recovering the chemical compound "X."

- The cost-benefit analysis was performed in advance;
- The *kaikaku* project was considered strategic–environmental concerns being of major importance at "JJ-Plant";
- The MCPD project team has carefully followed the manufacturing improvement budgets cycle (the five steps described in Section 3.3.1).

Step 4: Action Plan for MCI: From Setting the CCLW Target to Implementing a Consistent Solution

- A team of five members and one project leader was set up;
- The four-week deadline was set to identify improvements and another eight weeks given to fully implement the chosen solutions;
- The main tasks of the team were:
  - To perform an assessment of the current state of Process 5;
  - To check the current status of CLW associated to environmental costs and check the current level of the CCLW declared in the system;
  - To prepare a plan of consistent measures to eliminate environmental costs;
  - To propose countermeasures for issues identified in Process 5;
  - To define new working standards for Process 5 in connection with chemical compound "X."

7.2.3 Phase 3: Manufacturing Cost Policy Management

Step 5: Engage the Workforce for MCI: Analyzing and Identifying Improvement Solutions to Reach the CCLW Target

- The team analyzed:
  - The layout of Process 5;
  - All equipment parameters of Process 5 for each production area: temperature, pressure, cycle time, process lead time, etc.;
  - The principles of achieving the six major types of products in Process 5;
  - The number of people on the three shifts;
  - The volume of production per shift;
  - The consumption of chemical compound "X" per shift by product type;

- Scrap rate;
- Aspects of safety and health related to chemical compound "X."
- The team analyzed the types of activities in which the chemical compound "X" is used and the daily average consumption of the chemical compound "X" by type of activity and in combination with other chemicals;
- The identified solution: acquiring and installing an installation to recycle chemical compound "X" within the company;
- The team received and analyzed offers from two companies manufacturing recycling systems for chemical compound "X";
- Feasibility studies have been conducted for both offers (the technical criterion was the capacity of the recycling tank; the economic criterion was the cost per 100 kg of recycling per day);
- One supplier was chosen from the two who fulfilled the technical and economic criteria and who was willing to make two changes to the recycling system: (1) changing the distillation system and (2) modifying the heat exchanger.

Step 6: MCI Performance: Checking the CCLW Improvements Results

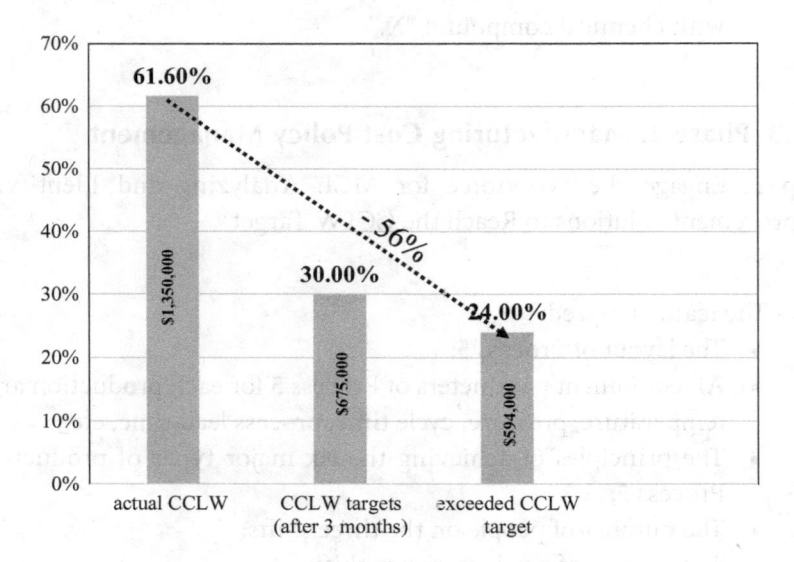

FIGURE 7.8

Annual overrun CCLW target for reducing environmental costs (by recovering the chemical compound "X").

- The CCLW target for reducing environmental costs by recycling the chemical compound "X" was reached within 12 weeks of the *kaizen* project for MCI being launched;
- The $675,000 target was met and exceeded (see Figure 7.8);
- The MCPD project team has carried out the performance analysis of manufacturing improvement budget targets and the actual manufacturing budgets drawn up; the targets have been met;
- The team has carried out the analysis of the performance of the cash flow improvement budget targets and the actual cash budgets drawn up; the targets have been met;
- The MCPD project team carried out the evaluation of the PFC losses and waste performance indicators after implementing CCLW and CLW improvements (after 30 days, following MCI management process–see Section 3.4.3);
- The MCPD team leader of this project has conducted an evaluation of the degree of involvement of team members to achieve the MCI target of this project.

Step 7: Daily MCI Management: Standardization and Expansion of Validated Solutions

- The team has developed and implemented a new standard for the use of chemical compound "X";
- The MCPD team leader of this project has continuously updated the state of the MCI improvement project at the MCPD system information center.

Future activities:

- It is planned to use this approach to Process 2 for chemical compound "Y."

Conclusion

Therefore, this MCI systemic improvement project, along with other MCI improvement projects, has met the annual MCI goal at "JJ-Plant." "JJ-Plant" has introduced a series of other systemic structural change projects such as: 5S expansion in all areas of the company, operative maintenance, preventive maintenance, quality assurance, redesigning the cost and budgeting system. "JJ-Plant" has a symbiosis between the traditional budgeting and costing system and the AMIB (for existing and new products) and AMCIB.

7.3 IMPROVEMENT PROJECT 11: SYSTEMIC MCI BY LABOR COST REDUCTION

Company background:

- "LL-Plant" joined the MCPD system two years ago;
- Type of industry: manufacturing and assembly–metal products;
- Manufacturing regime: repeated lot–machine shop, press lines and assembly lines;
- PFC number: 2;
- Improvement area: line 5–PFC 2 (motion losses).

Defining the problem: increasing the productivity of workers by introducing simple automation device–labor cost reduction.

7.3.1 Phase 1: Manufacturing Cost Policy Analysis

Step 1: From Company Productivity Vision to Establish Annual MCI Goal: Current State of the Process/Equipment from the CCLW Perspective

- Annual MCI goal: $3,200,000;
- Number of MCI means (annual strategic improvement projects to meet annual MCI goal by CAMPT): 14 *kaizen* projects and 3 *kaikaku* projects–this project is one of these three (similar to Figure 3.13);
- Stake to improve CLW associated with human work losses–motion losses: 11% of annual MCI goal.

Step 2: Target Setting for CCLW: Full Understanding of the Current State of Losses and Waste

- After completing and carefully analyzing the 7 matrices (for PFC analyzed), annual CCLW related with human work losses for line 5 (motion losses): $69,000;

- The theme of the *kaizen* project for human work losses for line 5 (motion losses): increase in capacity in manual assembly operation 16 in line 5;
- Annual MCI targets for human work losses for line 5–operation 16 (motion losses): human work–improving motion losses; the need to reduce/eliminate CCLW coordinates with the need for improvement (see Figure 7.9);
- The current capacity of support activity of operation 16 of line 5 is 1,116 products per shift (for the 3 shifts);
- The required capacity is 2,391 products per shift.

7.3.2 Phase 2: Manufacturing Cost Policy Development

Step 3: Annual Budgets for MCI: The Impact of CCLW Improvement through This Project from the Annual MCI Goal Perspective

- The cost-benefit analysis was performed in advance; the cost associated with the improvement was insignificant; the improvement costs were mainly associated with the *kaikaku* team's hourly cost and the manufacture of a simple automation device–the device was developed within the "LL-Plant";

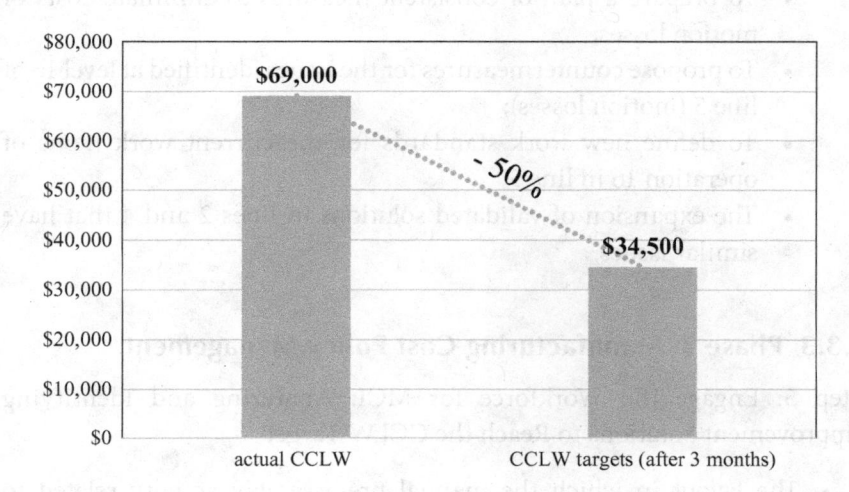

FIGURE 7.9
Setting annual CCLW target for human work losses for line 5–operation 16 (motion losses).

- The profitability of the lost opportunity to achieve production was analyzed;
- The MCPD project team has carefully followed the manufacturing improvement budgets cycle (the five steps described in Section 3.3.1).

Step 4: Action Plan for MCI: From Setting the CCLW Target to Implementing the New Improvement Standard

- A *kaikaku* team consisting of six members and one project leader was established;
- The 4-week deadline was set to identify improvements and another 8 weeks given to fully implement the new standard of work;
- The main tasks of the team were:
  - To perform an assessment of the current state of operation 16 in line 5 (motion losses)
  - To check the current status of the CLW associated with operation 16 in line 5 (costs of motion losses), and check the current level of the CCLW declared in the system ($69,000; the amount was correct; the manual assembly activity behind operation 16 on line 5 sometimes had a lead time that was not synchronized to the takt time of line 5; sometimes this activity became a bottleneck);
  - To prepare a plan of consistent measures to eliminate costs of motion losses;
  - To propose countermeasures for the issues identified at level 16 of line 5 (motion losses);
  - To define new work standards for the current work items of operation 16 in line 5;
  - The expansion of validated solutions in lines 2 and 4 that have similar issues.

7.3.3 Phase 3: Manufacturing Cost Policy Management

Step 5: Engage the Workforce for MCI: Analyzing and Identifying Improvement Solutions to Reach the CCLW Target

- The layout in which the manual pre-assembly activity related to operation 16 of line 5 was performed has been analyzed;

- The team analyzed steps of the manual assembly process of the subassembly for "EE" component (support activity of operation 16 in line 5; two subassemblies for a product).
- Manual assembly of "EE" was carried out by two operators;
- The team measured the operator time cycle for the manual assembly of "EE" (30 seconds in total):
  - Cycle time for operator 1: 14 seconds;
  - Cycle time for operator 2: 16 seconds;
- The current capacity of the two operators was 2,232 manual assemblies per shift for 1,116 replacement products (66,960 seconds available per shift/30 seconds of total cycle time/2);
- Required capacity 2,391 products per shift (66,960 seconds available per shift/14 seconds total cycle time / 2);
- Following the total cycle time and process lead time for support analysis of operation 16 in line 5, the following were identified: 2 work items could be combined and one work item could be eliminated by introducing a simple automation device (simple "EE" assembly automation);
- Description of the automation device:
  - Assembling the three parts on a pneumatic device with automatic exhaust of assembled part;
  - The mechanism transformed the input and movement forces by moving components (belts and rollers) into a desired set of forces and output movements;
  - Reduction of travel and multiple manipulation of people were targeted without connecting this mechanism to a computer;
  - Mechanism control was given by the design of the mechanics (gravitational energy is transformed in motion - based on the weight of the goods);
  - Reduction/elimination improvements have been made for: Component feed; operator movements; packaging time; transfer time; positioning times; insertion times.

Step 6: MCI Performance: Checking the CCLW Improvements Results

- The level of CCLW for human work losses for line 5 operation 16 (motion losses) reached 13 seconds in 7 weeks of the start of the *kaizen* project for MCI;

- Annual MCI targets for equipment cycle time reduction was fully met ($34,500);
- The new "EE" assembly capacity is 2,690 products per shift achieved by a single operator (see Figure 7.10);
- The MCPD project team has carried out the performance analysis of manufacturing improvement budget targets and the actual manufacturing budgets drawn up; the targets have been met;
- The team has carried out the analysis of the performance of the cash flow improvement budget targets and the actual cash budgets drawn up; the targets have been met;
- The MCPD project team carried out the evaluation of the PFC losses and waste performance indicators after implementing CCLW and CLW improvements (after 30 days, following MCI management process–see Section 3.4.3);
- The MCPD team leader of this project has conducted an evaluation of the degree of involvement of team members to achieve the MCI target of this project.

Step 7: Daily MCI Management: Standardization and Expansion of Validated Solutions

- The team has developed and implemented the SOP for a single cycle time of 13 seconds for an "EE" assembly.

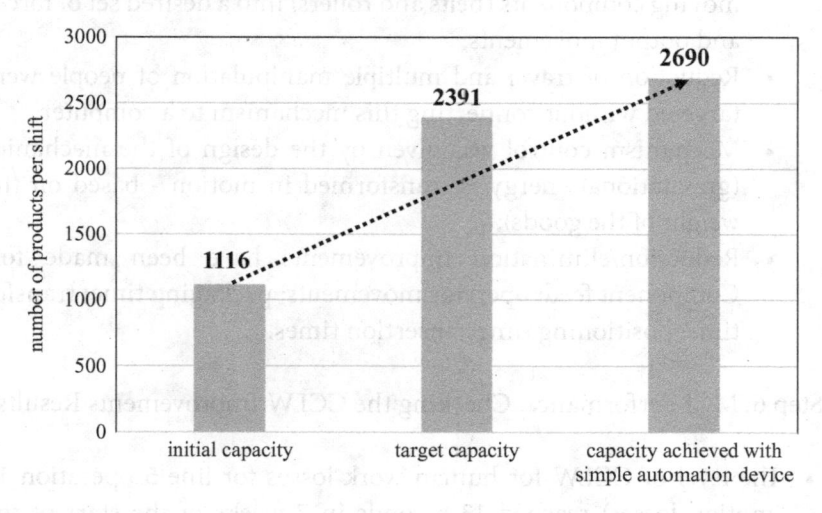

FIGURE 7.10
Fulfilling and overtaking the target capacity with the simple automation device for line 5–operation 16 (motion losses).

Future activities:

- The extension of improvement (simple automation device) for lines 2 and 4 has been planned.

This chapter outlined three systemic improvement projects (*kaikaku*) of CCLW to meet the annual MCI target, given that the companies were in need of minimizing inputs by increasing manufacturing control amid the very slow growth of the number of products to be sold or amid a decrease in sales (see Figure 3.2). Improvements targeted SDI (2)–CLW 4, 6 and 7 (see Matrices 1–7 from Figures 3.4, 3.9, 3.11 and 3.12).

8

MCI through Aligning the Factory Lead Time and Cycle Time to Takt Time

Continuous improvement of company efficiency by continually improving manufacturing lead time alignment to takt time is a basic concern of MCI change drivers on the need to increase manufacturing delivery (minimizing inputs—"I") to support external manufacturing target profits and, especially, internal manufacturing target profits (see Figure 3.2).

Continuous knowledge of LW level and of CLW associate factory lead time (Posteucă and Zapciu, 2015) and, in particular, associate supply chain lead time (waste), manufacturing lead time (waste), delivery lead time (waste) and continuous reduction of CCLW by planning and running the required annual MCI means on time is required.

Four systematic improvement projects (*kaizen*) will be presented in this chapter. Section 8.1 presents a *kaizen* project for MCI by reducing the lead time of an assembly line. Section 8.2 presents a *kaizen* project for MCI by increasing the synchronization level of the equipment lead time with the takt time of the assembly month. Section 8.3 presents a *kaizen* project for the reduction of WIP storage space. Finally, Section 8.4 presents a *kaizen* project for MCI for reducing unplanned packaging consumption and reducing or eliminating supply lead time.

Therefore, all of these *kaizen* projects for MCI target SDI (3)–CLW 9–14 improvements (see Matrices 1–7 from Figures 3.4, 3.9, 3.11 and 3.12).

Setting annual MCI targets for CLW 9–14 is carried out to increase the number of products manufactured and sold by improving manufacturing delivery (see Figure 8.1).

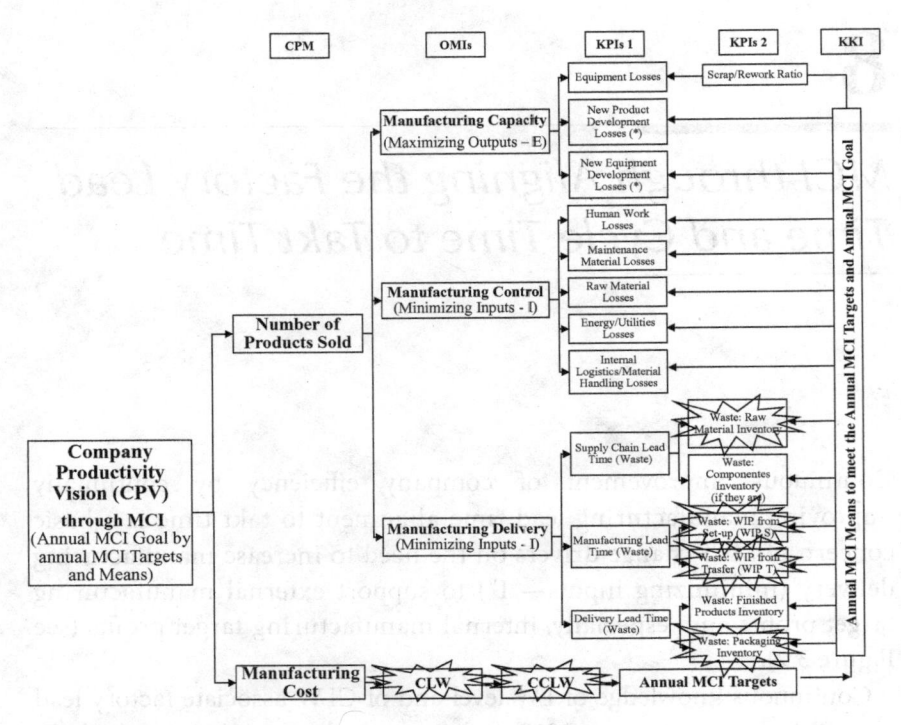

FIGURE 8.1
MCI policy deployment for a PFC for manufacturing delivery.

8.1 IMPROVEMENT PROJECT 12: SYSTEMATIC MCI BY REDUCING THE LEAD TIME OF AN ASSEMBLY LINE

Company background:

- "PP-Plant" joined the MCPD system three years ago;
- Type of industry: manufacturing and assembly–components for the automotive industry;
- Manufacturing regime: repeated lot–machine shop, press lines and assembly lines;
- PFC number: 2;
- Improvement area: assembly lines (line 2);
- The set-up time at line 2 had been improved a short while ago and is considered to be an acceptable time for that given moment; therefore, improving lead time on line 2 will not include the set-up of 570 seconds;

- A complex *kaizen* project was designed at line 2 a short while ago to improve line capacity by improving: (1) breakdowns; (2) scrap and rework; (3) minor stoppages; (4) cycle time–line balancing based on takt time and ergonomics; and (5) shutdown–especially line stoppages due to shortage of timely supplies;
- A *kaizen* project was designed at line 2 a short while ago for optimizing the route and loading/unloading of the line feed train (internal logistics).

8.1.1 Phase 1: Manufacturing Cost Policy Analysis

Step 1: From Company Productivity Vision to Establish Annual MCI Goal: Current State of Line 2 From the CCW Perspective

- Annual MCI goal: $14,500,000;
- Number of MCI means (annual strategic improvement projects to meet annual MCI goal by CAMPT): 15 *kaizen* projects and 3 *kaikaku* projects (similar to Figure 3.13);
- Stake to improve CLW associated with line 2 lead time reduction: 6% of annual MCI goal (sixth place as weight; 1 *kaizen* strategic project was selected out of 15).

Step 2: Target Setting for CCW: Full Understanding of the Current State of Waste

- Annual CCW related with line 2 lead time reduction: $875,000; cost of waste is calculated based on the annual lead time for line 2 and the impact of line 2 on the entire manufacturing flow for PFC 2 (the seconds along the entire manufacturing flow affected by the lead time of line 2 have been calculated–before and after line 2, and lead time in line 2; the seconds are transformed into WIP T/S; WIP T/S is converted into equivalent finished products; the opportunity for lost profitability was established for finished products);
- The theme of the *kaizen* project: line 2 lead time reduction (one-piece-flow for line 2);
- Line 2 is part of PFC 2;
- Line length: 44 meters;
- Annual MCI target for line 2 lead time reduction (see Figure 8.2);

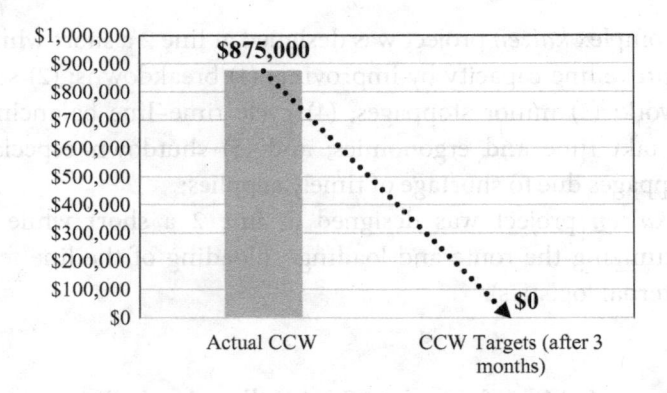

FIGURE 8.2
Setting annual CCW target for line 2 lead time reduction.

- Annual MCI means target for line 2 lead time reduction to meet annual CCW target (see Figure 8.3); WIP T: 57.5 finished product (cycle time: 26 seconds).

8.1.2 Phase 2: Manufacturing Cost Policy Development

Step 3: Annual Budgets for MCI: The Impact of CCW Improvement through This Project from the Annual MCI Goal Perspective

- The cost-benefit analysis was performed in advance;
- The cost associated with the improvement was insignificant; the costs of improvement were mainly associated with the *kaizen* team's hourly cost;

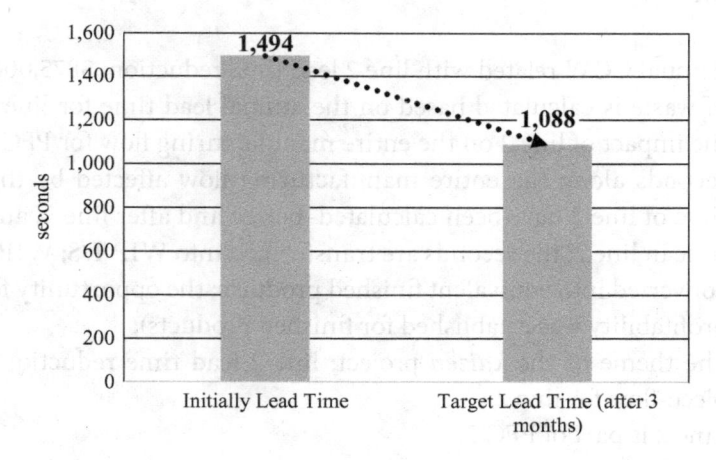

FIGURE 8.3
Set the target of reducing the line 2 lead time.

- The profitability of the lost opportunity to achieve production was analyzed for line 2 lead time reduction;
- The MCPD project team has carefully followed the manufacturing improvement budgets cycle (the five steps described in Section 3.3.1).

Step 4: Action Plan for MCI: From Setting the CCW Target to Implement a New Lead Time on Line 2

- A team consisting of four members and one project leader was established;
- The 3-week deadline was set to identify improvements and another 8 weeks given for implementation, verification and standardization of new SOPs;
- The main tasks of the team were:
  - To perform an assessment of the current state of each area of line 2;
  - To check the current status of CCW associated with line 2;
  - To conceive a plan of consistent measures to reduce lead time at line 2;
  - To propose countermeasures for the problems identified in each area of line 2 where the lead time is high;
  - To define new standards of work for areas where lead time will be reduced.

8.1.3 Phase 3: Manufacturing Cost Policy Management

Step 5: Engage the Workforce for MCI: Analyzing and Identifying Improvement Solutions to Reach the CCW Target

- The team analyzed the layout of line 2;
- The team measured the current lead time of line 2 for representative products;
- The team measured the lead time of the previous processes of line 2 and of processes after line 2 to analyze the framing of each lead time in the current and future takt time;
- The team measured set-up time at line 2 for representative products; the average time set-up measured was 570 seconds or 9.5 minutes;

- The team measured the lead time at line 2 for representative products; the lead time measured coincided with the one in the system (1,494 seconds, 24.9 minutes–set-up time at line 2 of 570 is included in 1,494 seconds);
- Therefore, the possible lead time to improve was 924 seconds or 15.4 minutes;
- Assembly line 2 had 7 defined areas;
- The team measured the lead time for each of the 7 areas and cycle time for each operation in the 7 areas;
- The longest lead time was identified in:
  - Area 4 (assembly 2): 334 seconds (an unnecessary space was identified–no operation was carried out there);
    - WIP T: 9 finished products (cycle time: 26 seconds);
    - Unnecessary line length in Area 4: 5 meters;
    - The opportunities for improving the lead time in Area 4 were analyzed (synonymous with Figure 5.10–this time for lead time).
  - Area 6 (functional test area): 442 seconds;
    - WIP T: 17 finished products (cycle time: 26 seconds);
    - Lead time with transfer to functional testing of products: 11 seconds (no operations are performed–rotary table);
    - Functional test zone: 431 seconds; 4 operators; 4 types of testing;
    - The opportunities to improve lead time in Area 6 were analyzed.
- New working methods have been identified for Areas 4 and 6, following ECRS analysis for basic function (BF) and especially for auxiliary function (AF), based on target setting for lead time improvement and for achieving the CCW target (with the help of *kaizenshiro*) (Posteucă and Sakamoto, 2017, pp. 327–352);
- The solutions identified were:
  - For Area 4: reducing the line by five meters and rebalancing operations;
  - For Area 6: layout redesign and simplification of testing operations (especially post-testing operations and up to Area 7; some of the equipment underwent adjustments: transfer chain, functional test carousel, rotary table).

Step 6: MCI Performance: Checking the CCW Improvements Results

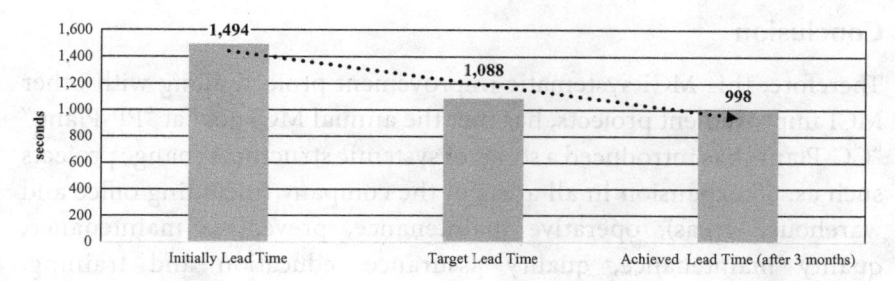

FIGURE 8.4
Fulfillment of lead time reduction to line 2: initial status versus target versus achievement.

- CCW level for line 2 lead time reduction (one-piece-flow for line 2) reached 528 seconds in the 11 weeks since the start of the *kaizen* project for MCI;
- Annual MCI targets for equipment cycle time reduction was fully achieved ($875,000)
- Lead time achieved in Area 4: 32 seconds;
- Lead time achieved in Area 6: 248 seconds;
- The new lead time for line 2 (one-piece-flow for line 2): 998 seconds (see Figure 8.4);
- Line 2 length after improvement: 39 meters (a 5-meter reduction);
- WIP T saved: 19 finished products (1,494 seconds reduced to 998 seconds; cycle time: 26 seconds);
- The MCPD team leader of this project has conducted an evaluation of the degree of involvement of team members to achieve the MCI target of this project.

Step 7: Daily MCI Management: Standardization and Expansion of Validated Solutions

- The team has developed and implemented a new standard for Areas 4 and 6 of line 2;
- The MCPD team leader of this project has continuously updated the state of the MCI improvement project at the MCPD system information center.

Future activities:

- Lead time analysis at line 1 was planned.

Conclusion

Therefore, this MCI systematic improvement project, along with other MCI improvement projects, has met the annual MCI goal at "PP-Plant." "CC-Plant" has introduced a series of systemic structural change projects such as: 5S expansion in all areas of the company (including office and warehouse areas), operative maintenance, preventive maintenance, quality maintenance, quality assurance, education and training, redesigning the cost and budget system to capture the variations between the target cost that has been set and the actual cost that has been achieved and the differences between the target cost and the zero cost of losses and waste. PP-Plant continuously monitors the price level of PFC-related products, the required unitary profit level, the manufacturing unit cost required to be reached and the CLW and CCLW level for each PFC. "PP-Plant" has a system of prioritization in choosing systematic and systemic strategic annual improvement projects to achieve annual MCI goal. "PP-Plant" has a symbiosis between the traditional budgeting and costing system and the AMIB (for existing and new products) and AMCIB. The top management at "PP-Plant" keeps track of what is the normal cost reduction potential for each PFC. In this way, top management consistently supports vision, mission and strategies of productivity in the short term and especially in the medium and long term. Finally, daily MCI management ensures an effective and efficiency dialogue between all levels of the company.

8.2 IMPROVEMENT PROJECT 13: SYSTEMATIC MCI BY SYNCHRONIZING THE LEAD TIME OF THE EQUIPMENT WITH THE ASSEMBLY LINE TAKT TIME

Company background:

- "LL-Plant" joined the MCPD system two years ago;
- Type of industry: manufacturing and assembly–components for the automotive industry;
- Manufacturing regime: repeated lot–machine shop, press lines and assembly lines;
- PFC number: 2;
- Improvement area: "C" equipment and assembly line 2.

8.2.1 Phase 1: Manufacturing Cost Policy Analysis

Step 1: From Company Productivity Vision to Establish Annual MCI Goal: Synchronizing the Lead Time of Equipment "C" with the Assembly Line 2 Takt Time from the CCLW Perspective

- Annual MCI goal: $3,500,000;
- Number of MCI means (annual strategic improvement projects to meet annual MCI goal by CAMPT): 10 *kaizen* projects and 2 *kaikaku* project (similar to Figure 3.13);
- Stake to improve CLW associated with synchronizing the lean time of equipment "C" with the assembly line 2 takt time: 8% of annual MCI goal (ninth place as weight; one *kaizen* strategic project was selected out of ten).

Step 2: Target Setting for CCLW: Full Understanding of the Current State of Assembly Line 2 Losses and Equipment Waste

- After completing and carefully analyzing the 7 matrices (for PFC analyzed), annual CCLW related with synchronizing the lead time of equipment "C" with the assembly line 2 takt time: $275,000;
- The theme of the *kaizen* project was to synchronize the "C" equipment with the assembly line 2 takt time: capacity increase for assembly line 2;
- Information on line 2 capacity affected by equipment "C":
  - Number of events per year (missing landmarks from equipment "C" at assembly line 2): 75 events;
  - Average event duration: 56.5 minutes;
  - Total duration of minutes per year: 4,240 minutes (254,400 seconds per year);
  - Lost production opportunity: 9,784 assembled finished products (takt time of the line is 26 seconds); or in other words: 3.8 shifts per year with no activity load due to lack of benchmarks from equipment "C" at assembly line 2 (average number of seconds available per shift: 66,860 seconds; a line OEE of 85%).
- Annual MCI targets for synchronizing the "C" equipment with the assembly line 2 takt time: capacity increase for assembly line 2 (reducing shutdown losses on line 2–lack of components); the need to reduce/eliminate CCLW coordinates the need for improvement;

the amount of $275,000 includes CLWs for the 4,240 minutes of line 2 and CLW from the processes affected by shutdown losses on line 2–lack of components (see Figure 8.5);

• Annual MCI means targets for shutdown losses on line 2–lack of components (see Figure 8.6).

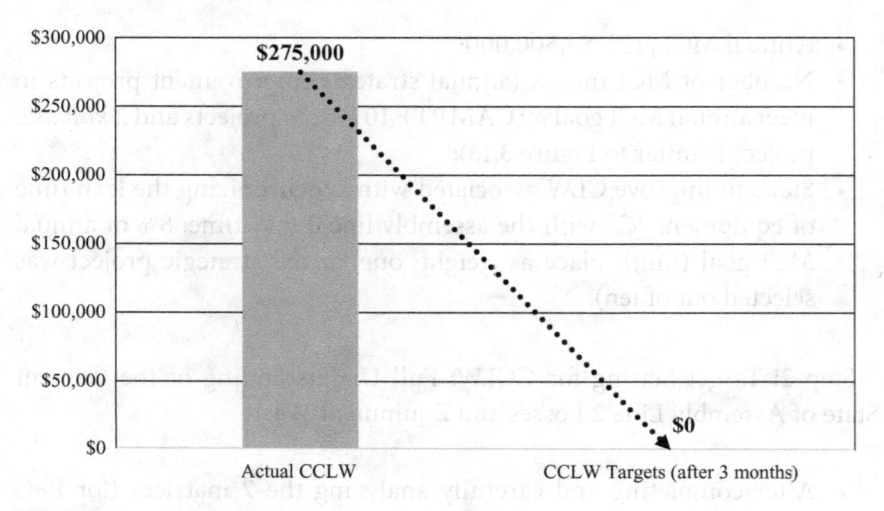

FIGURE 8.5
Setting annual CCW target for reducing shutdown losses on line 2–lack of components from "C" equipment.

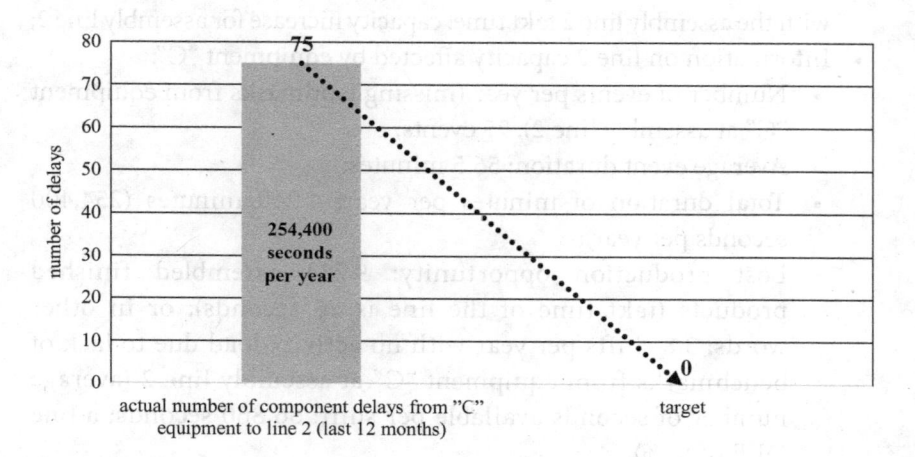

FIGURE 8.6
Set the target to reduce the number of delays for the timely delivery of components from the "C" equipment to the assembly line 2.

8.2.2 Phase 2: Manufacturing Cost Policy Development

Step 3: Annual Budgets for MCI: The Impact of CCLW Improvement through This Project from the Annual MCI Goal Perspective

- The cost-benefit analysis was performed in advance;
- The cost associated with the improvement was insignificant (software developed as improvement was created by IT specialists within "PP-Plant");
- Improvement costs were mainly associated with the *kaizen* team hourly cost;
- The profitability of the lost opportunity to achieve production was analyzed;
- The MCPD project team has carefully followed the manufacturing improvement budgets cycle (the five steps described in Section 3.3.1).

Step 4: Action Plan for MCI: From Setting the CCLW Target to Implementing the New Improvement Standard

- A team consisting of five members and one project leader was established;
- A 5-week deadline was set to identify improvements and another maximum of 8 weeks given to fully implement the improvements (the new standard);
- The main tasks of the team were:
  - To make an assessment of the current status of the equipment "C";
  - To check the current status of CLW associated to equipment "C" losses and the internal logistics from equipment "C" to line 2 (lead time); verifying the current level of CCLW declared in the system;
  - To conceive a plan of consistent measures to eliminate component delays from the equipment "C" to the assembly line 2;
  - To propose countermeasures for the problems identified in the equipment "C" level and at the level of internal logistics from the equipment "C" to line 2 (lead time);
  - To define new work standards (SOPs);
  - To extend validated solutions to other similar equipment and to line 1.

8.2.3 Phase 3: Manufacturing Cost Policy Management

Step 5: Engage the Workforce for MCI: Analyzing and Identifying Improvement Solutions to Reach the CCLW Target

- The team analyzed the layout of the equipment "C" area, the associated technological flow and the route traveled by the components manufactured at equipment "C" for line 2;
- The benchmarks created by equipment "C" for "line 2" have been identified;
- The parameters of equipment were analyzed (they were within standard limits);
- The temporary storage areas of the equipment "C" benchmarks were analyzed (next to equipment "C," in the warehouse and near line 2; the minimum and maximum level practiced versus the declared one in the system was checked);
- All production planning involving equipment "C" and line 2 was analyzed in detail;
- The equipment "C" was a unique plant at "PP-Plant";
- The following major issues have been identified:
  - The lack of real-time benchmark stock control of the equipment "C";
  - The lack of WIP control (between equipment "C" and assembly line 2);
  - Unacceptably large variation in the buffer stock level of equipment "C" (to cover the lack of temporary capacity of the equipment "C");
  - The lack of strict control of the stock outflows for the benchmarks of the equipment "C";
  - The lack of accurate and timely information on the level of rework and scrap for line 2;
  - The lack of WIP synchronization of the equipment "C" with set-up time;
  - The lack of WIP synchronization related to the equipment "C" benchmarks with the production plan changes.
- The seven issues identified above have been analyzed in detail; the main phenomena have been defined as follows: (1) inventory management is flawed; (2) line 2 does not communicate production plan changes in due time; and (3) line 2 does not

communicate early rework and scrap and set-up planning in due time. Equipment "C" has the capability to meet the volumes required by line 2;

- The improvement proposed: creating software for line 2 computers and two computers from the "C" equipment (located one at the input and the other at the output of the equipment "C" process) in order to achieve real-time stock management and to achieve continuous communication with line 2;

- The inventory level is not done manually at the end of each shift, but is done with real-time software (real-time control of inputs and outputs of the inventory "C" of all benchmarks manufactured by equipment "C," including for line 2; the traceability of the equipment "C" was accurate after software implementation);

- The parts are planned and made in a synchronized way with the production lines (including line 2), and the hourly control of the orders and the stock of benchmarks from the equipment "C" is performed;

- The new software performs production planning for the equipment "C" and sets production priorities for the assembly lines (flexibility of the equipment "C" is good; the set-up time is 11.5 minutes);

- The new software allows highlighting scrap and rework in real time from each line;

- No major internal logistics problems were identified (lead time for the train).

Step 6: MCI Performance: Checking the CCLW Improvements Results

- The CCLW level for synchronizing the lead time of the equipment "C" with the assembly line 2 takt time has completely eliminated 75 delays in 10 weeks since the *kaizen* project for MCI started;

- Annual MCI targets for synchronizing the lead time of equipment "C" with the assembly line 2 takt time has been fully achieved ($275,000);

- The lost production opportunity of 9,784 finished products assembled per year on line 2 has been eliminated completely;

- The MCPD project team has carried out the performance analysis of manufacturing improvement budget targets and the actual manufacturing budgets drawn up; the targets have been met;

- The team has performed the analysis of the performance of the cash flow improvement budget targets and the actual cash budgets drawn up; the targets have been met;
- The MCPD project team carried out the evaluation of the PFC losses and waste performance indicators after implementing CCLW and CLW improvements (after 30 days, following MCI management process–see Section 3.4.3);
- The MCPD team leader of this project has conducted an evaluation of the degree of involvement of team members to achieve the MCI target of this project.

Step 7: Daily MCI Management: Standardization and Expansion of Validated Solutions

- The IT team has developed and implemented the new software to achieve synchronization of the lead time of equipment "C" with the assembly line 2 takt time;
- The *kaizen* project team has developed the new SOPs to implement the new software;
- The MCPD team leader of this project has continuously updated the state of the MCI improvement project at the MCPD system information center.

Future activities:

- Expansion of the software made for equipment "C" to eight other equipment was planned, with the ultimate goal being the total traceability of the traceability of all raw materials and components for each product.

Conclusion

Therefore, this MCI systematic improvement project, along with other MCI improvement projects, has met the annual MCI goal at "LL-Plant." "LL-Plant" has introduced a series of systemic structural change projects such as: 5S expansion in all areas of the company (including office and warehouse areas), operative maintenance, preventive maintenance, quality maintenance, quality assurance, education and training, redesigning the cost and budgeting system.

8.3 IMPROVEMENT PROJECT 14: SYSTEMATIC MCI BY REDUCTION OF WIP STORAGE SPACE

Company background:

- "RR-Plant" joined the MCPD system 2.5 years ago;
- Type of industry: process–food;
- Manufacturing regime: batches–rolling packaging;
- PFC number: 1;
- Improvement area: storage space of WIP for food semi-processing.

8.3.1 Phase 1: Manufacturing Cost Policy Analysis

Step 1: From Company Productivity Vision to Establish Annual MCI Goal: Current State of the Storage Space of WIP for Food Semi-Processing from the CCLW Perspective

- Annual company MCI goal: $3,500,000;
- Number of annual company MCI means (annual strategic improvement projects to meet annual MCI goal by CAMPT): 16 *kaizen* projects;
- Stake to improve CLW associated with storage space of WIP for food semi-processing: 1.8% of annual MCI goal.

Step 2: Target Setting for CCLW: Full Understanding of the Current State of Waste

- After completing and carefully analyzing the 7 matrices (for PFC analyzed), annual CCLW related with storage space of WIP for food semi-processing: $65,000;
- The theme of the *kaizen* project for reduction of WIP storage space: optimizing the loading of shelves in the warehouse with semi-finished products ("SD");
- Current information about the "SD" semi-processed food:
  - It is performed in Area 3 of "RR-Plant";
  - The number of pieces of similar equipment that produce the "SD" semi-processed food: 12 pieces of equipment;
  - Equipment cycle time: 60 seconds;

- Storage capacity: 69 locations in the warehouse (207 pallets);
- Number of operators: 9 operators per shift;
- Number of shifts: 3;
- After processing of the "SD" semi-finished product, it is packed in plastic bags placed on pallets; pallets are transported in the WIP semi-finished product warehouse;
- The average use of space in the WIP warehouse allocated to the "SD" semi-finished product is only 58% (even if a minimum and maximum use of each storage cell was defined, the cells were not occupied at their maximum capacity);
- The total volume of a standard location is 6.9 cubic meters; the average of the occupied volume was 4 cubic meters; the average of the unoccupied volume was 2.9 cubic meters;
- Long feed times with the "SD" semi-finished product of the processes manufacturing the final product products (unnecessary routes in the WIP warehouse for the "SD" semi-finished product);
- Annual MCI targets for reduction of WIP storage space for semi-finished products ("SD") (see Figure 8.7);
- Annual MCI means targets for reduction of WIP storage space for semi-finished products ("SD") (see Figure 8.8); the target setting has taken into account both the production plan for the next period and the need for annual MCI targets.

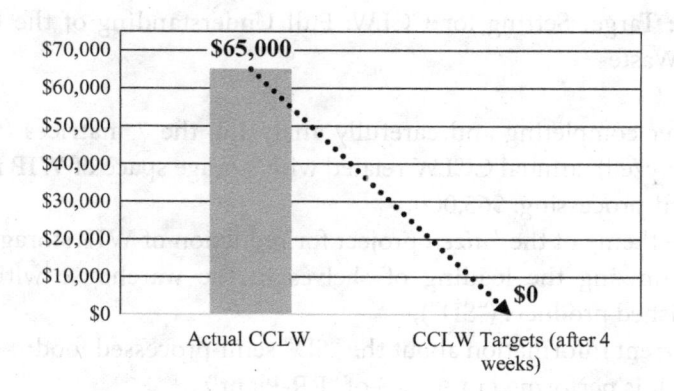

FIGURE 8.7
Setting annual CCW target for reduction of WIP storage space for semi-finished products ("SD").

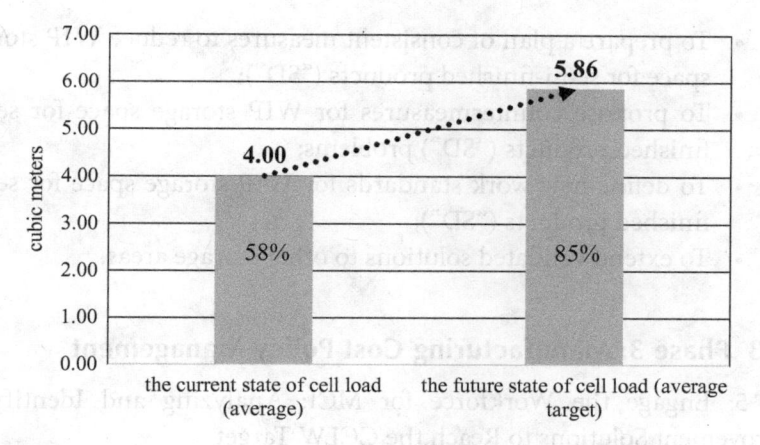

FIGURE 8.8
Setting the target to increase the shelf load from the warehouse with semi-finished products ("SD").

8.3.2 Phase 2: Manufacturing Cost Policy Development

Step 3: Annual Budgets for MCI: The Impact of CCLW Improvement through This Project from the Annual MCI Goal Perspective

- The cost-benefit analysis was performed in advance;
- The cost associated with the improvement was insignificant;
- Improvement costs were mainly associated with the *kaizen* team's hourly cost;
- The MCPD project team has carefully followed the manufacturing improvement budget cycle (the five steps described in Section 3.3.1).

Step 4: Action Plan for MCI: From Setting the CCLW Target to Implementing the New Improvement Standard

- A team consisting of four members and one project leader was established;
- A 4-week deadline has been set to fully implement the new standard;
- The main tasks of the team were:
  - To perform an assessment of the current status of storage for "SD";
  - To check the current status of the CLW associated with "SD" storage and check the current level of CCLW declared in the system;

- To prepare a plan of consistent measures to reduce WIP storage space for semi-finished products ("SD");
- To propose countermeasures for WIP storage space for semi-finished products ("SD") problems;
- To define new work standards for WIP storage space for semi-finished products ("SD");
- To extend validated solutions to other storage areas.

8.3.3 Phase 3: Manufacturing Cost Policy Management

Step 5: Engage the Workforce for MCI: Analyzing and Identifying Improvement Solutions to Reach the CCLW Target

- The team analyzed layout in Area 3 (the 10 machines that were also manufacturing the "SD" among other things), the layout of the storage areas for "SD" and the layout of the finishing-product areas where "SD" was brought;
- The team measured the current location of each location (the 69 locations in the warehouse; 207 pallets) and set their loading average of 58%;
- The team checked the production plan for the next 12 months and determined the need to increase storage capacity to 85%;
- The team proposed changing the storage method: from plastic bags to boxes;
- The team proposed to use a single box size–six boxes on a pallet.

Step 6: MCI Performance: Checking the CCLW Improvements Results

- The storage loading level reached 88% (beyond the 85% target);
- The CCW level related with storage space of WIP for the semi-processed food was reached within 4 weeks of the start of the *kaizen* project for MCI (target of $65,000);
- The feeding times with the "SD" semi-finished product of the manufacturing area decreased by 75% (from 450 seconds to 112 seconds);
- The number of cells allocated for "SD" in the warehouse decreased: from 69 to 35;
- The number of handling for the return of "SD" remaining stocks to the finished product area decreased;

- The stock level accuracy for "SD" increased (reducing the differences between the physical stock and the accounting stock - in the system);
- FIFO has been better observed;
- A series of packing materials such as stretch film have been eliminated;
- The MCPD team leader of this project has conducted an evaluation of the degree of involvement of team members to achieve the MCI target of this project.

Step 7: Daily MCI Management: Standardization and Expansion of Validated Solutions

- The team has developed and implemented a new standard for storage space of WIP for "SD" semi-finished products;
- The MCPD team leader of this project has continuously updated the state of the MCI improvement project at the MCPD system information center.

Future activities:

- Expansion of the improvement to another type of semi-finished products has been planned.

Conclusion

Therefore, this MCI systematic improvement project, along with other MCI improvement projects, has met the annual MCI goal at "RR-Plant." "RR-Plant" continuously monitors the price level of PFC-related products, the required unitary profit level, the unit cost required to be reached and the CLW and CCLW level for each PFC. "RR-Plant" has a system of prioritization in choosing systematic and systemic strategic annual improvement projects to achieve annual MCI goal. "RR-Plant" has a symbiosis between the traditional budgeting and costing system and the AMIB (for existing and new products) and AMCIB. Top management at "RR-Plant" keeps track of what is the normal cost reduction potential for each PFC and each product by continually measuring CLW, by highlighting CLW and CCLW in company's internal financial statements and by strategic planning of annual MCI means. In this way, the top management consistently supports vision, mission and strategies of productivity in the short term and especially in the medium and long term.

8.4 IMPROVEMENT PROJECT 15: SYSTEMATIC MCI BY REDUCING UNPLANNED PACKAGING CONSUMPTION TO IMPROVE SUPPLIER LEAD TIME

Company background:

- "TT-Plant" joined the MCPD system 1.5 years ago;
- Type of industry: process–pharmaceuticals;
- Manufacturing regime: batches–rolling packaging;
- PFC number: 6;
- Improvement area: packaging and storage of finished products.

8.4.1 Phase 1: Manufacturing Cost Policy Analysis

Step 1: From Company Productivity Vision to Establish Annual MCI Goal

- Annual MCI goal: $5,000,000;
- Number of MCI means (annual strategic improvement projects to meet annual MCI goal by CAMPT): 12 *kaizen* projects and 2 *kaikaku* project;
- Stake to improve CLW associated with unplanned consumption with air bubble roll stretch: 3.5% of annual MCI goal.

Step 2: Target Setting for CCW: Full Understanding of the Current State of Waste

- Standard annual consumption for air bubble roll stretch: $1,400,000;
- After completing and carefully analyzing the 7 matrices (for PFC analyzed), annual CCW related with unplanned consumption with air bubble roll stretch (consumption over the annual standard): $350,000;
- Types of air bubble roll stretch used: 3 types (Type 1: 97%);
- Problems caused by unplanned air bubble roll stretch consumption:
  - Problems in calculating the budget and standard cost of products;
  - Urgent orders from air bubble roll stretch providers (not observing the standard lead time);
  - Stock differences between the actual stock and the accounting stock.

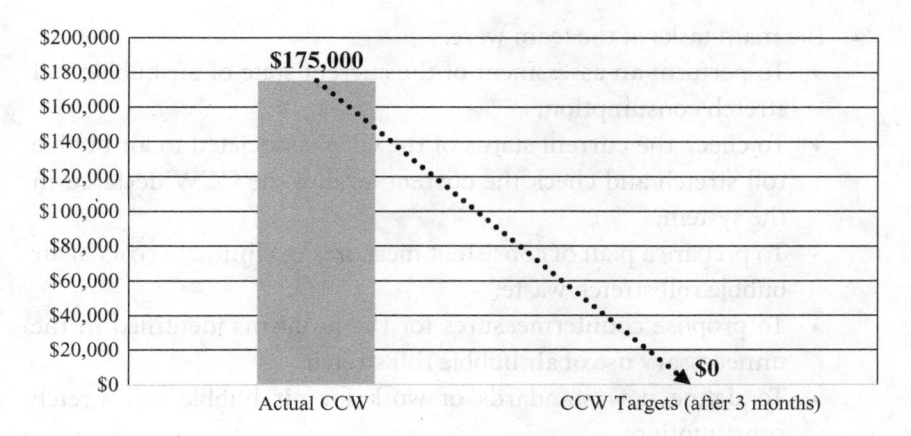

FIGURE 8.9
Setting annual CCW target for reduction of unplanned consumption with air bubble roll stretch.

- The theme of the *kaizen* project: reducing unplanned consumption with air bubble roll stretch;
- Annual CCW target related with unplanned consumption with air bubble roll stretch (50% reduction of current consumption over standard): $175,000 (see Figure 8.9).

8.4.2 Phase 2: Manufacturing Cost Policy Development

Step 3: Annual Budgets for MCI: The Impact of CCLW Improvement through This Project from the Annual MCI Goal Perspective

- The cost associated with the improvement was insignificant;
- The improvement costs were mainly associated with the *kaizen* team's hourly cost;
- The MCPD project team has carefully followed the manufacturing improvement budgets cycle (the five steps described in Section 3.3.1).

Step 4: Action Plan for MCI: From Setting the CCLW Target to Implementing the New Improvement Standard

- A team consisting of seven members and a project leader was established;
- A 3-month deadline has been set to identify improvements and implement them fully (the new standard);

- The main tasks of the team were:
  - To perform an assessment of the current state of air bubble roll stretch consumption;
  - To check the current status of the CLW associated to air bubble roll stretch and check the current level of the CCW declared in the system;
  - To prepare a plan of consistent measures to eliminate costs of air bubble roll stretch waste;
  - To propose countermeasures for the problems identified in the unnecessary use of air bubble roll stretch;
  - To define new standards of work for air bubble roll stretch consumption;
  - To extend validated solutions to other types of packaging materials.

8.4.3 Phase 3: Manufacturing Cost Policy Management

Step 5: Engage the Workforce for MCI: Analyzing and Identifying Improvement Solutions to Reach the CCW Target

- The team analyzed the layout to identify areas where air bubble roll stretch is consumed; four areas were identified (predominantly Type 1 air bubble roll stretch);
- The team analyzed:
  - The types of products that used air bubble roll stretch when packaged;
  - The department responsible for the air bubble roll stretch for each of the four areas;
  - The declared standard consumption;
  - The actual consumption registered in the system;
  - The unplanned air bubble roll stretch consumption for the past 12 months for each area and for each product type.
- The team analyzed the causes of unplanned air bubble roll stretch consumption and identified the following:
  - The operators did not observe standard consumption because they had not attended the training (using more air bubble roll stretch than required);
  - The operators did not observe the standard consumption and used more air bubble roll stretch to avoid some quality problems

that were known at their level but were not analyzed at higher levels;

- Using air bubble roll stretch for other types of packing that did not require this type of material.
- The team reviewed all air bubble roll stretch standards for all product types (packing standards);
- The team has identified new packaging methods that will reduce the unplanned air bubble roll stretch consumption;
- The operators participated in the air bubble roll stretch training for the packaging of the main products (for all SOPs);
- The team proposed the use of a cheaper air bubble roll stretch (for Type 2 and Type 3 air bubble roll stretch).

Step 6: MCI Performance: Checking the CCW Improvements Results

- The level of CCW associated with unplanned air bubble roll stretch was achieved and exceeded in the 3 months allocated;
- Annual MCI targets for reducing unplanned consumption with air bubble roll stretch was fully met ($192,500 instead of $175,000; a decrease by 55%);
- The MCPD project team has carried out the performance analysis of manufacturing improvement budget targets and the actual manufacturing budgets drawn up; the targets have been met;
- The team has carried out the analysis of the performance of the cash flow improvement budget targets and the actual cash budgets drawn up; the targets have been met;
- The MCPD project team carried out the evaluation of the PFC losses and waste performance indicators after implementing CCLW and CLW improvements (after 30 days, following MCI management process–see Section 3.4.3);
- The MCPD team leader of this project has conducted an evaluation of the degree of involvement of team members to achieve the MCI target of this project.

Step 7: Daily MCI Management: Standardization and Expansion of Validated Solutions

- The team has developed and implemented 17 new packaging standards using air bubble roll stretch;

- The MCPD team leader of this project has continuously updated the state of the MCI improvement project at the MCPD system information center.

Future activities:

- A new *kaizen* project was decided upon to reduce the air bubble roll stretch consumption by at least 90%;
- Three other *kaizen* projects were planned to reduce three other types of packaging material consumption.

Four CCLW systematic (*kaizen*) projects were presented in this chapter to meet the annual MCI goal, given that the companies were in need of minimizing inputs by increasing manufacturing delivery amid the very slow growth of the number of products to be sold or due to a decrease in sales (see Figure 3.2). Improvements targeted SDI (3)–CLW 9, 11, 12 and 14 (see Matrices 1–7 from Figures 3.4, 3.9, 3.11 and 3.12).

Appendix 1: The MCPD Transformation Checklist

The 100 questions of the MCPD transformation checklist are the list of priorities and tasks of all people involved in the implementation, maintenance and continuous development of the MCPD system in line with the company productivity vision (CPV).

The MCPD transformation checklist is a quick way to determine the progress of manufacturing flow transformation for each PFC through the MCPD system. At the same time, using the MCPD transformation checklist, it is easier to integrate any new member of a manufacturing company's management team.

Each MCPD transformation checklist, as an evolutionary process, can be tailored to the current and future specificity and situation of the manufacturing company and is extremely useful at least in the early years of implementing the MCPD system.

Note: in our example, the manufacturing company structure consists of:

1. Board of Directors
2. Senior Managers
3. Plant Manager
4. Marketing
5. Sales and Research
6. Production
7. Maintenance
8. Production Engineering
9. Quality Assurance
10. Production Control
11. Supply Chain
12. Human Resources
13. Continuous Improvement
14. Research and Development
15. Sourcing, Resource Management
16. Managerial and Cost Accounting

TABLE A.1

The MCPD Transformation Checklist

The MCPD Transformation Checklist		The rating is scoring from 1 to 5 for MCPD activity and company structure (1 = MIN; 5 = MAX.)																	Date:			
PFC name to be evaluated:																			Test No.:			
Members of the MCPD team:		Manufacturing Company Structure															Evaluation		Countermeasures			
Activities	Questions	1	2	3	4	5	6	7	8	9	10	11	12	13	14	15	16	Actual	Target	Actions	Start	End
Introduction	Has a steering committee of MCPD been formed?	X	X	5	X	X	5	5	3	4	4	2	3	5	3	3	5	47	60			
	Have the company productivity vision and mission been established and disseminated in the company on hierarchical levels?	5	5	5	5	5	5	5	4	4	3	5	4	4	3	4	4	70	80			
	Have productivity core business goals been set?	X	X	5	X	X	5	4	3	4	4	3	4	4	2	2	4	44	60			
	Have multiannual basic productivity strategies been established?	X	5	4	X	X	4	5	5	4	3	4	2	3	2	4	4	49	65			
	Has a productivity master plan been achieved?	X	X	5	X	X	5	4	4	4	4	3	3	3	3	4	4	46	60			
	Have the seven core principles of the MCPD system been internalized by top management?	3	5	5	5	5	5	5	5	5	5	4	5	5	4	3	4	73	80			

(Continued)

	Is the practical production capacity known by top management?	5	5	5	X	X	5	4	4	4	5	4	4	4	4	4	5	62	70
	Has a cost-saving strategy been objectively set for each product family cost?	X	5	5	X	X	4	5	5	4	4	3	3	3	3	4	5	53	65
	Does each hierarchical level exactly know the tasks from the perspective of the MCPD system?	5	5	3	5	5	5	5	5	5	5	3	4	4	3	4	5	71	80
	Actual (amounted)	18	25	42	15	15	43	42	38	38	36	32	32	34	26	33	41	510	620
	Target (maximum)	20	25	45	15	15	45	45	45	45	45	45	45	45	45	45	45	615	620
	Percentage of achievement for each company structure (%)	90	100	93	100	100	96	93	84	84	80	71	71	76	58	73	91	85.069	1
	Overall rating (A— excellent: 80%–100%, B—good: 70%–80%, C—unsatisfactory: less than 70%)	A	A	A	A	A	A	A	A	A	A	B	B	B	C	C	A		
	Result on manufacturing company structures (%)					85.069									A				
Background and business needs	Have OMIs been set for the next year?																		

(Continued)

TABLE A.1 (CONTINUED)

The MCPD Transformation Checklist

Was an annual external and internal target profit set for each PFC at the profit and loss statement (P&L) level?

Was an annual MCI goal set for each product family cost?

Are we exactly and continuously aware of the cost per minute of equipment running in each cost center for each PFC?

Are we exactly and consistently aware of the cost of one minute of direct and indirect labor from each cost center for each PFC?

Are the latest price benchmark analysis results known?

Is the exact price trend for each competitor product known?

(Continued)

TABLE A.1 (CONTINUED)

The MCPD Transformation Checklist

	Actual (amounted)
	Target (maximum)
	Percentage of achievement for each company structure (%)
	Overall rating (A—excellent: 80%–100%, B—good: 70%–80%, C—unsatisfactory: less than 70%)
	Result on manufacturing company structures (%)
Establishment of MCPD System	Have the processes of each PFC been defined and selected?
	Has company productivity policy deployment been established?
	Has MCI policy deployment for losses and waste been established?
	Has MCI policy deployment for CLW been established?

(Continued)

TABLE A.1 (CONTINUED)

The MCPD Transformation Checklist

Does the top management know exactly the state of scenarios hypostases (growth or decline in sales) for each PFC?
Are the eight change drivers of MCI known by top management at the level of each process of each PFC?
Has each manager his set of KPIs, with updated, referring to MCI?
Is each manager responsible for annual MCI means (*kaizen* and *kaikaku* for MCI)?
Have annual MCPD implementation cross-functional teams been formed?
Has an annual internal and external communication plan related to MCPD system goals been achieved?

(Continued)

	Actual (amounted)
	Target (maximum)
	Percentage of achievement for each company structure (%)
	Overall rating (A—excellent: 80%–100%, B—good: 70%–80%, C—unsatisfactory: less than 70%)
	Result on manufacturing company structures (%)
Step 1: Context and Purpose of MCI (Plan)	Have strategic key points on manufacturing processes been defined and identified?
	Have productivity and manufacturing cost improvement strategies been set?
	Has annual productivity policy deployment been achieved for each PFC?

(Continued)

TABLE A.1 (CONTINUED)

The MCPD Transformation Checklist

Has annual MCI policy deployment been achieved to establish and annual MCI goal?
Is the annual MCI goal objective and sufficient to meet annual manufacturing target profit?
Are system elements of change drivers of MCI known for each PFC?
Is the need to reduce manufacturing unit costs known in dynamics for every product of each product family?
Actual (amounted)
Target (maximum)
Percentage of achievement for each company structure (%)
Overall rating (A—excellent: 80%–100%, B—good: 70%–80%, C—unsatisfactory: less than 70%)

(Continued)

The MCPD Transformation Checklist

	Result on manufacturing company structures (%)
Step 2: Annual MCI Targets and Means (Plan)	Was Matrix 1: continuous measurement of losses and waste drawn up?
	Was Matrix 2: analyze and identify sources that generate losses and waste drawn up?
	Was Matrix 3: continually converting losses and waste into manufacturing costs drawn up?
	Was Matrix 4: analyze and establish critical costs of losses and waste drawn up?
	Was Matrix 5: current assumptions for critical costs of losses and waste improvement drawn up?
	Was Matrix 6: setting annual MCI targets to achieve annual MCI goals drawn up?

(*Continued*)

TABLE A.1 (CONTINUED)

The MCPD Transformation Checklist

	Was Matrix 7: setting annual MCI means to achieve annual MCI targets drawn up?
	Has cause–effect analysis been conducted along with PFC processes for losses and waste behavior?
	Has annual and monthly valuation of CLW been made from total manufacturing costs for the main processes of PFC?
	Has aligning directions for annual strategic direction of improvement been made based on product life cycle sales scenarios for each PFC?
	Is the difference between actual CLW and ideal costs known continuously for each PFC?
	Is the difference between actual CLW and ideal costs known continuously for each process of each PFC?

(Continued)

The MCPD Transformation Checklist

Is the difference between actual CLW and future costs known continuously for each PFC?
Is the difference between actual CLW and future costs known continuously for each process of each PFC?
Has the annual strategic direction of improvement been defined for each PFC?
Was the critical to annual manufacturing profitability tree drawn up?
Was a reconciling top-down and bottom-up for annual MCI goal drawn up for each PFC – based on catchball process?
Actual (amounted)
Target (maximum)
Percentage of achievement for each company structure (%)

(*Continued*)

TABLE A.1 (CONTINUED)

The MCPD Transformation Checklist

	Overall rating (A—excellent: 80%–100%, B—good: 70%–80%, C—unsatisfactory: less than 70%)	
	Result on manufacturing company structures (%)	
Step 3: Annual Budgets for MCI (Plan)	Have fixed manufacturing CLW targets been established based on catchball process?	
	Have variable manufacturing CLW targets been established based on catchball process?	
	Have fixed manufacturing CCLW targets been established based on catchball process?	
	Have variable manufacturing CCLW targets been established based on catchball process?	

(Continued)

TABLE A.1 (CONTINUED)

The MCPD Transformation Checklist

Have CLW targets been established for general and administrative expenses based on catchball process?	
Have CCLW targets been established for general and administrative expenses based on catchball process?	
Have manufacturing cash improvement budget targets been established based on catchball process?	
Have annual manufacturing improvement budgets been made for existing products?	
Have annual manufacturing improvement budgets been made for future products?	
Has an annual manufacturing cash improvement budget been drawn up?	

(Continued)

TABLE A.1 (CONTINUED)

The MCPD Transformation Checklist

Has a performance analysis between manufacturing improvement budgets targets and actual manufacturing budgets been drawn up?
Has a performance analysis between manufacturing cash improvement budgets targets and actual cash budgets drawn up?
Are continuous monitoring and evaluation of manufacturing improvement budgets made?
Has the feasibility for each *kaizen* and *kaikaku* project for MCI been made?
Actual (amounted)
Target (maximum)
Percentage of achievement for each company structure (%)

(Continued)

TABLE A.1 (CONTINUED)

The MCPD Transformation Checklist

	Overall rating (A—excellent: 80%–100%, B—good: 70%–80%, C—unsatisfactory: less than 70%)	
	Result on manufacturing company structures (%)	
Step 4: Action Plan for MCI (Plan)	Have *kaizen* and *kaikaku* annual strategic projects been defined for each PFC?	
	Has a list of all MCI actions and activities been prepared for each PFC?	
	Has annual planning of the annual *kaizen* and *kaikaku* strategic projects for MCI been performed?	
	Have the individual plans for MCI been defined and planned?	
	Are the results of annual MCI means (*kaizen* and *kaikaku*) being monitored and evaluated?	

(Continued)

TABLE A.1 (CONTINUED)

The MCPD Transformation Checklist

Are used versus planned resources being monitored and evaluated for each *kaizen* and *kaikaku* project?
Are the gained versus planned benefits being monitored and evaluated for each *kaizen* and *kaikaku* project?
Is the status of each *kaizen* and *kaikaku* project being monitored and evaluated?
Are the benefits gained over time for *kaizen* and *kaikaku* projects being monitored and evaluated?
Actual (amounted)
Target (maximum)
Percentage of achievement for each company structure (%)
Overall rating (A— excellent: 80%–100%, B—good: 70%–80%, C—unsatisfactory: less than 70%)

TABLE A.1 (CONTINUED)

The MCPD Transformation Checklist

	Result on manufacturing company structures (%)
Step 5: Engage the Workforce for MCI (Do)	Is interdepartmental organization fully operational to ensure meeting annual MCI targets?
	Have all training needs been set up to meet the annual MCI means targets?
	Have all workshop needs been set up to meet annual MCI means targets?
	Is the annual training and workshop plan continually updated to meet annual MCI means targets?
	Is the annual training and workshop plan achieved for all hierarchical levels to meet annual MCI means targets?
	Is each *kaizen* and *kaikaku* project following strictly the seven steps of the MCPD system?

TABLE A.1 (CONTINUED)

The MCPD Transformation Checklist

	Are *kaizen* and *kaikaku* projects for MCI a habit?
	Actual (amounted)
	Target (maximum)
	Percentage of achievement for each company structure (%)
	Overall rating (A— excellent: 80%–100%, B—good: 70%–80%, C—unsatisfactory: less than 70%)
	Result on manufacturing company structures (%)
Step 6: MCI Performance Management (Check)	Is AMIB/AMCIB monthly performance assessment performed for each PFC?
	Is AMIB/AMCIB for each PFC adjusted for every 6 months?
	Is monthly assessment of the degree of employee involvement performed to achieve MCI targets (for each manufacturing area)?

(Continued)

Is annual assessment of systemic cost improvement performance management performed?

Is monthly assessment of KPIs related to losses and waste performed?

Is monthly assessment of KPIs related to CLW performed?

Is monthly assessment of the achieved level of annual MCI goal performed?

Is monthly assessment of meeting annual MCI means targets performed?

Is monthly assessment by departments of the participation in the fulfillment of the annual MCI targets and means performed?

Are 6-month and 12-month assessments of the intangible effects of implementing the MCPD system performed?

(Continued)

TABLE A.1 (CONTINUED)

The MCPD Transformation Checklist

	Are 3-month, 6-month and 12-month assessments of the consistency of the implemented solutions performed?
	Actual (amounted)
	Target (maximum)
	Percentage of achievement for each company structure (%)
	Overall rating (A—excellent: 80%–100%, B—good: 70%–80%, C—unsatisfactory: less than 70%)
	Result on manufacturing company structures (%)
Step 7: Daily MCI Management (Act)	Are the tangible effects of KPIs related to MCI monitored daily?
	Have control items for annual MCI means targets (for each manufacturing area) been defined?

(Continued)

Have control points for annual MCI means targets (for each manufacturing area) been defined?

Have inspection items for annual MCI means targets (for each manufacturing area) been defined?

Have inspection points f annual MCI means targets (for each manufacturing area) been defined?

Are KPIs related to MCI analyzed daily in connection with other KPIs of manufacturing company (for each manufacturing area)?

Is the daily MCI management process (for each manufacturing area) respected?

Is the MCPD system information center updated daily?

(Continued)

TABLE A.1 (CONTINUED)

The MCPD Transformation Checklist

Is management branding monitored daily and evaluated every 6 and 12 months?
Actual (amounted)
Target (maximum)
Percentage of achievement for each company structure (%)
Overall rating (A—excellent: 80%–100%, B—good: 70%–80%, C—unsatisfactory: less than 70%)
Result on manufacturing company structures (%)

Appendix 2: Glossary—The Basic Concepts of the MCPD System

Over the years, the author has received several requests from the main audience for the MCPD system—top management, middle management and professional support staff who manage manufacturing areas—to define the usual terms of the MCPD system, from *cost of losses and waste* to *zero costs of losses and waste*. Moreover, some of the discussions with the MCPD practitioners in the manufacturing company have revealed some small confusions and inconsistent use of terms such as *annual MCI means targets* (sometimes confused with *annual MCI targets*).

In the following glossary, one can find the explanation of the basic terms of the MCPD system to help both practitioners and theorists fully understand the MCPD system. Any suggestion to improve the clarity of the terms described below is welcome (these can be sent to: office@exegens. com). At the same time, these terms will be placed on the Exegens® website to be reviewed in future publications of the MCPD system (https:// exegens.com).

Annual Manufacturing Cash Improvement Budget (AMCIB) is an element of the third step of the MCPD system. It is an extension of the annual manufacturing improvement budget (AMIB) for existing and new products and an extension of the company's conventional cash flow. The role of AMCIB is to support the annual stake of MCPD for all product family costs of the annual MCI goal, by planning and evaluating in money and gains obtained as a result of meeting the annual MCI means targets.

Annual Manufacturing Improvement Budgets (AMIB) is an element of the third step of the MCPD system. They are the approach of all PFC of the company with the aim of planning and controlling the level of the annual MCI goal through the development of: (1) annual manufacturing improvement budget for existing products, (2) multiannual manufacturing improvement budget for new products.

Critical to Annual Manufacturing Profitability Tree (CAMPT) is the tool for selecting, planning and targeting all *kaizen* and *kaikaku* improvement projects to achieve annual MCI goal.

Critical Costs of Losses and Waste (CCLW): see Matrix 4 (Analyze and Establish Critical Costs of Losses and Waste).

Costs of Losses and Waste (CLW): see Matrix 3 (Continually Converting Losses and Waste into Manufacturing Costs).

Company Productivity Mission (CPM) is the company productivity vision (CPV) extension and is the second step of the productivity business model (PBM)—the MCPD system framework. It is the manufacturing capacity required to be provided by the company manufacturing system for the next 5–10 years and is the task of senior managers.

Company Productivity Strategy (CPS) is the extension of productivity core business goals (PCBG) and is the fourth step of the productivity business model (PBM)–the MCPD system framework. It represents the detail of the PCBG goals and directions and how the basic productivity strategy, the departmental productivity strategy and the family product productivity strategy will contribute to the achievement of annual and multiannual manufacturing profit targets. It is the task of senior and middle managers.

Company Productivity Vision (CPV) sets out the visionary directions to make profit through productivity and is the first step of the productivity business model (PBM)—the MCPD system framework. It is the Board of Directors' statement on the number of products to be sold and the profit for 5–10 years for each PFC.

Daily Cost Management (DCM) is the seventh step of the MCPD system (daily MCI Management). It refers to day-to-day control of processes and equipment, or shop floor management for MCI, to verify productivity productivity at all KPIs, and for verifying the fulfillment degree of MCI policy deployment for all PFCs from the KPIs related to losses and waste.

Daily Manufacturing Cost Improvement Process (DMCIP) is the basic tool of the seventh step of the MCPD system (daily MCI management). It is a set of principles, processes and tools that enable monitoring and achieving annual MCI targets and means at all levels of the organization and clearly establish decision-making levels by designating owners for all KPIs related to MCI ensuring fast horizontal and vertical communication to continually have a quick reaction and an adequate response to any problem with the annual MCI means targets.

Daily Management Indicators (DMIs) are the extension of the *kaizen* and *kaikaku* indicators (KKIs) for MCI and are an element of the fifth step of the productivity business model (PBM)—the MCPD system framework. They identify, report and address deviations from standard as quickly as possible for company generic KPIs (**production, quality, cost, delivery, safety, morale** (PQCDSM)), for KPIs related to losses and waste, CLW and CCLW and for KKIs through day-to-day problem-solving activities, through the participation of team leaders and operators.

External Manufacturing Target Profit is the multiannual and annual manufacturing target profit obtained from manufacturing and selling products to customers. Calculation: multiannual and annual manufacturing target profit–multiannual and annual external manufacturing profit=multiannual and annual internal manufacturing target profit.

Internal Manufacturing Target Profit is the multiannual and annual manufacturing target profit obtained from MCI. Calculation: multiannual and annual manufacturing target profit–multiannual and annual external manufacturing profit=multiannual and annual internal manufacturing target profit.

***Kaizen* and *Kaikaku* Indicators (KKIs)** are the key performance indicators (KPIs) extension and are an element of the fifth step of the productivity business model (PBM)—the MCPD system framework. They are continually assessing the success of the KPIs related to losses and waste, CLW and CCLW at the level of each PFC process.

Losses and Waste Stratification Flow Analysis (LWSFA) is the measurement of the impact of losses or waste (cause) on all losses and waste affected from upstream processes (effects), from downstream processes (effects) and from the process in which losses or waste occurred (effects). The MCPD team seeks to identify as many possible effects of a cause as possible to determine CCLW (see Matrix 4).

Management Branding (MB) "is a managerial system that, by an integrated approach, creates and synchronizes, for the application, contextual managerial behavioral identities in order to increase organizational productivity and/or economic growth." (Posteucă, 2011).

Multiannual Basic Productivity Strategies (MBPS) is a tool to achieve multiannual manufacturing profits for: (1) ensuring an acceptable level of effectiveness in achieving the annual and multiannual target number of units to be produced and sold (by maximizing outputs; reducing not effectively used input–losses improvement); (2) to ensure an acceptable level of efficiency to meet annual and multiannual manufacturing target cost (by minimizing inputs; reducing excess amount of input–waste improvement).

Management by Manufacturing Cost Improvement Policy (MMCIP) is the management of the fulfillment of annual and multiannual manufacturing costs and targets.

Manufacturing Cost Improvement (MCI) it is the process of searching, identifying and removing CLW and CCLW from the processes of each PFC by setting targets and means to continually target ideal costs (zero CLW).

Manufacturing Cost Improvement Catchball Process (MCICP) it is the process of participatory decision-making on annual MCI targets and means through which information and ideas are thrown and caught, back and forth, up and down throughout the manufacturing company for each PFC.

Manufacturing Cost Policy analysis (MCPa) is the first phase of the MCPD system—the first two steps of the MCPD system. It represents the process to translate the need to reduce the manufacturing unit costs into concrete actions and activities at the level of the main processes of each PFC, and to increase real managerial commitment and reduce resistance to change among all people in the company and beyond.

Manufacturing Cost Policy development (MCPd) is the second phase of the MCPD system—the third and fourth steps of the MCPD system. It is the process of coordinating the MCI means to achieve MCI targets in order to consistently support MCPD in the medium-term and the long-term and to ensure horizontal and vertical communication through the total involvement of all departments in order to achieve the continuous transformation of the company by reducing reactive managerial behavior to operational challenges.

Manufacturing Cost Policy management (MCPm) is the third phase of the MCPD system—the fifth, sixth and seventh steps of the MCPD system (last steps). It represents the process of total involvement of all the people in the company and beyond to fulfill the MCI means, monitoring, assessing and adjusting the effectiveness of the MCPD system.

Matrix 1 of MCPD: Continuous Measurement of Losses and Waste is part of the second step of the MCPD system. It is part of the ongoing and consistent reconciliation process between the top-down and bottom-up approach for establishing the annual MCI target for each PFC. It is the qualitative analysis of identifying the opportunities for consistent MCI approach to losses and waste for each major process of each PFC and manufacturing system with KPIs for losses and waste (the level of non-productivity is measured continuously and consistently in the last 6–12 months).

Matrix 2 of MCPD: Analyze and Identify Sources that Generate Losses and Waste is part of the second step of the MCPD system. It is part of the ongoing and consistent reconciliation process between the top-down and bottom-up approach for establishing the annual MCI target for each PFC. It is performing an analysis of causes of losses and waste at their root where they are formed, and then preventing them from appearing and having effects at the level of each PFC and entire manufacturing system (ragged effects most often in inventory and stocks).

Matrix 3 of MCPD: Continually Converting Losses and Waste into Manufacturing Costs is part of the second step of the MCPD system. It is part of the ongoing and consistent reconciliation process between the top-down and bottom-up approach for establishing the annual MCI target for each PFC. It is a quantitative analysis to identify the CLW in a proportion of 30–40% of current manufacturing costs for each PFC in order to: (1) determinate the total MCI multiannual reserve; (2) establish the basis for annual CCLW and implicitly for annual MCI targets and means; (3) establish the annual and multiannual cost reduction plan for the CLW for each cost object (such as: CLW for each product; CLW for all PFC products and CLW for a process/equipment/cost center).

Matrix 4 of MCPD: Analyze and Establish Critical Costs of Losses and Waste is part of the second step of the MCPD system. It is part of the ongoing and consistent reconciliation process between the top-down and bottom-up approach for establishing the annual MCI target for each PFC. It is the process of determining the acceptable future status of the total CLW of each PFC and total company for the next five years and aims at identifying CCLW that is at least 80% of the CLW identified by Matrix 3.

Matrix 5 of MCPD: Current Assumptions for Critical Costs of Losses and Waste Improvement is part of the second step of the MCPD system. It is a part of the ongoing and consistent reconciliation process between the top-down and bottom-up approach for establishing the annual MCI target for each PFC. It is the process of determining the annual level of CLW that can be addressed by reducing the annual CCLW level for each PFC. It seeks to identify and consistently address the main phenomena, principles and symptoms of CLW by developing current assumptions for

CCLW, especially credible assumptions, after the TOCR has been set for the next five years to reduce CLW.

Matrix 6 of MCPD: Setting Annual MCI Targets to Achieve Annual MCI Goal is part of the second step of the MCPD system. It is part of the ongoing and consistent reconciliation process between the top-down and bottom-up approach for establishing the annual MCI target for each PFC. It is the process of setting the annual MCI targets that converge to meet the annual MCI goal. Matrix 6 develops together with Matrix 5 and Matrix 7 with CAMPT and annual budgets for MCI, after the current assumptions for CCLW improvement for the next year have been established.

Matrix 7 of MCPD: Setting Annual MCI Means to Achieve Annual MCI Targets is a part of the second step of the MCPD system. It is a part of the ongoing and consistent reconciliation process between the top-down and bottom-up approach for establishing the annual MCI target for each PFC. It is the process of establishing the link between annual MCI means and annual MCI goals through the fulfillment of annual basic MCI strategies.

Manufacturing Cost Policy Deployment is "… the process of translating the strategic objective of reducing manufacturing costs in the long run toward the improvement of annual systematic activities and toward annual systemic improvement actions by setting targets and means to improve process costs of families of products." (Posteucă, 2015, p. 65; Posteucă and Sakamoto, 2017, pp. 81–82; Posteucă, 2018, pp. 10–11).

Multiannual Manufacturing Improvement Budgets (MMIB) is the AMIB extension for current and future products of all PFCs of the company with the aim of planning and controlling the level of multiannual manufacturing profit targets (internal and external).

Productivity Business Model (PBM) is a process to support the full implementation of a robust multiannual productivity program at all levels of a company to support long-term profitability and competitiveness. It is the implementation framework of the MCPD system and consists of seven main phases: (I) Company's productivity vision (CPV); (II) Company's productivity mission (CPM); (III) Productivity core business goal strategies; (IV) Productivity Strategies (Level 1: the basic productivity strategies; Level 2: the departmental productivity strategies; and Level 3: the product family productivity strategy); (V) Productivity policy deployment (OMIs, KPIs, KKIs, DMIs); (VI) Productivity Master Plan (for five years, three years and one year, continuously updated for the next 12 months) and (VII) Productivity Continuous Feedback to CPV.

Productivity Core Business Goals (PCBG) is the company productivity missions (CPM) extension and is an element of the fifth step of the productivity business model (PBM)—the MCPD system framework. PCBG represents the major directions within the company, which needs to be continuously improved to meet the "productivity vision" based on the "productivity mission". PCBG highlights the key areas for which resources will be allocated in the future such as: to reduce the manufacturing delivery time; to reduce the manufacturing costs; to increase the manufacturing profit; to develop new products; to provide respect for people.

Productivity Master Plan (PMP) is the process of establishing an annual and especially multiannual (5 years) strategic productivity plan for each PFC.

Product Family Cost (PFC) is a product group that shares the same types of costs, equipment, and processing parameters. The definition of PFCs is made taking into account the following elements: (1) production numbers; (2) contribution to multiannual and annual target profits for the annual MCI target; (3) percentage of similar processes over the entire production stream; (4) products that have similar potential to reduce CLW and achieve the annual MCI targets; and (6) the product life cycle is increasing and/or falls within the initial volume and target profit (ultimately).

Physical Losses (PL) are the structures of losses and waste associated with material inputs from processes. The manufacturing cost is related to PL (or impact on manufacturing system inputs).

Strategic Direction of Improvement (SDI) represents the MCI policy deployment extensions for each PFC for strategically targeting productivity improvement needs. There are three types: SDI (1): Manufacturing Capacity (Maximizing Outputs "E"); SDI (2): Manufacturing Control (Minimizing Inputs "I"); and SDI (3) Manufacturing Delivery (Minimizing Inputs "I").

Strategic Key Points on Manufacturing Process (SKPMP) is the PCBG extension. It is the analysis for each PFC based on which the productivity strategies, annual productivity objectives, KPIs of CLW, *kaizen* and *kaikaku* activities and daily MCI management are developed at all hierarchical levels of the manufacturing company.

Time-Related Losses (TRL) are the structures of losses in process/equipment times. The transformation cost is related to TRL (or impact on manufacturing system outputs).

Total Offer for Cost Reduction (TOCR) is the acceptable future status of CLW for the next 5 years (targeting ideal costs zero CLW).

Zero Costs of Losses and Waste (ZCLW) is the ideal state or final destination of the successive transformation of the manufacturing flow through the continuous process of the seven steps of the MCPD (planning and execution of the future CLW/CCLW, usually annual) to reach ideal costs (zero CLW/CCLW) for each PFC to reduce/eliminate 30–40% of CLW in the total manufacturing costs of a PFC.

Bibliography

Ahuja, I. P. S. and Khamba, J. S., 2008. Total productive maintenance: Literature review and directions. *International Journal of Quality & Reliability Management*, 25(7), 709–756.

Akao, Y., 1991. *Hoshin Kanri: Policy Deployment for Successful TQM* (originally published as Hoshin Kanri Kaysuyo no jissai, 1988). New York, Productivity Press.

American National Standards Institute, 1983. *Industrial Engineering Terminology*. Hoboken, NJ: Wiley-Interscience.

Ansari, S., Bell, J. and Okano, H., 2006. Target costing: Uncharted research territory. In *Handbooks of Management Accounting Research*, vol. 2. Amsterdam, the Netherlands: Elsevier, pp. 507–530.

Bodek, N., 2004. *Kaikaku: The Power and Magic of Lean: A Study in Knowledge Transfer*. Vancouver, Washington: PCS Inc.

Burnham, D. C., 1972. *Productivity Improvement*. New York: Columbia University Press.

Burrows, G. and Chenhall, R. H., 2012. Target costing: First and second comings. *Accounting History Review*, 22(2), 127–142.

Cândido, C. J. and Santos, S. P., 2011. Is TQM more difficult to implement than other transformational strategies?. *Total Quality Management & Business Excellence*, 22(11), 1139–1164.

Carlson, C. R. and Wilmot, W. W., 2006. *Innovation: The Five Disciplines for Creating What Customers Want*. New York: Crown Business.

Chau, V. S. and Witcher, B. J., 2008. Dynamic capabilities for strategic team performance management: The case of Nissan. *Team Performance Management*, 14(3/4), 179–191.

Chiarini, A. and Vagnoni, E., 2014. World-class manufacturing by Fiat. Comparison with Toyota Production System from a Strategic Management, Management Accounting, Operations Management and Performance Measurement dimension. *International Journal of Production Research*, 53(2), 590–606.

Chiarini, A., 2016. Corporate social responsibility strategies using the TQM: Hoshin kanri as an alternative system to the balanced scorecard. *The TQM Journal*, 28(3), 360–376.

Choi, T. Y. and Liker, J. K., 1995. Bringing Japanese continuous improvement approaches to US manufacturing: The roles of process orientation and communications. *Decision Sciences*, 26(5), 589–620.

Cooper, R., 1995. *When Lean Enterprises Collide: Competing through Confrontation*. Boston, MA: Harvard Business Press.

Cooper, R., 1996. Costing techniques to support corporate strategy: Evidence from Japan, *Management Accounting Research*, 7(2), 219–246.

Cooper, R. and Slagmulder, R., 2004. Interorganizational cost management and relational context, *Accounting, Organizations and Society*, 29(1), 1–26.

Cooper, R. and Slagmulder, R., 1999. Developing profitable new products with target costing. *Sloan Management Review*, 40(4), 23–33.

Cua, K. O., McKone, K. E. and Schroeder, R. G., 2001. Relationships between implementation of TQM, JIT, and TPM and manufacturing performance, *Journal of Operations Management*, 19(6), 675–694.

Cudney, E. A., 2009. *Using Hoshin Kanri to Improve the Value Stream*. Boca Raton, FL: CRC Press.

Cudney, E. A., 2016. Development of strategic quality metrics for organizations using Hoshin Kanri. *Quality in the 21st Century*, UK: Springer.

Dahlgaard-Park, S. M., 2011. The quality movement: Where are you going?, *Total Quality Management & Business Excellence*, 22(5), 493–516.

Deming, W. E., 1986. *Out of the Crisis*. Cambridge, MA: Massachusetts Institute of Technology, Center for Advanced Engineering Study.

Drucker, P. F., 1963. *Managing for Business Effectiveness*. Boston, MA: Harvard University, Graduate School of Business Administration.

Drucker, P. F., 1973. *Management: Tasks, Responsibilities, Practices*. New York: Harper and Row.

Drucker, P. F., 2006. *The Practice of Management*, New York: HarperCollins.

Duarte, J. E., 1993. Policy deployment: Planning methods that get results. *CMA Magazine*, 67(4), 13.

Ellram, L. M., 2006. The implementation of target costing in the United States: Theory versus practice. *Journal of Supply Chain Management*, 42(1), 13–26.

Feigenbaum, A. V., 1956. Total quality-control. *Harvard Business Review*, 34(6), 93–101.

Gadiesh, O. and Gilbert, J. L., 1998. Profit Polls: A Fresh Look at Strategy. *Harvard Business Review*, May–June 1998.

Gåsvaer, D. and von Axelson, J., 2012. Kaikaku-radical improvement in production. In *International Conference on Operations and Maintenance*, Singapore, Singapore: World Academy of Science, Engineering and Technology, pp. 758–765.

Heap, J., 2008. Innovation and enterprise: The foundations of developing productivity. *International Journal of Productivity and Performance Management*, 57(6), 434–439.

Helmrich, K., 2003. *Productivity Process: Methods and Experiences of Measuring and Improving*. Stockholm, Sweden: International MTM Directorate.

Hino, S., 2006. *Inside the Mind of Toyota: Management Principles for Enduring Growth*. New York: Taylor & Francis Group.

Hirano, H., 2009. *JIT Implementation Manual - The Complete Guide to Just-in-Time Manufacturing*. New York: CRC Press.

Hosomi, S., Scarbrough, P. and Ueno, S., 2017. Management accounting in Japan: Current practices. *The Routledge Handbook of Accounting in Asia*, UK: Routledge.

Hutchins, D., 2008. *Hoshin Kanri: The Strategic Approach to Continuous Improvement*. Burlington, VA: Gower Publishing.

Imai, M. 1997. *Gemba Kaizen: A Commonsense Low—Cost Approach to Management*. New York: McGraw-Hill.

Ishikawa, K., 1980. *QC Circle Koryo: General Principles of the QC Circle*, Tokyo: QC Circle Headquarters, Union of Japanese Scientists and Engineers.

Jackson, T. L., 2006. *Hoshin Kanri for the Lean Enterprise: Developing Competitive Capabilities and Managing Profit*. New York: CRC Press.

Juran, J. M., 1995. *Managerial Breakthrough: The Classic Book on Improving Management Performance*. New York: McGraw-Hill.

Kaplan, R. S., 1990. *Measure for Manufacturing Excellence*. Boston, MA: Harvard Business School Press.

Kaplan, R. S. and Norton, D. P., 1996. *The Balanced Scorecard: Translating Strategy Into Action*. Boston, MA: Harvard Business School Press.

Kato, Y., 1993. Target costing support systems: Lessons from leading Japanese companies. *Management Accounting Research*, 4(1), 33–47.

Kondo, Y., 1998. Hoshin Kanri-a participative way of quality management in Japan. *The TQM Magazine*, 10(6), 425–431.

Lee, J. Y. and Monden, Y., 1996. Kaizen costing: Its structure and cost management functions. *Advances in Management Accounting*, 5, 27–40.

Lee, J. Y., Chen, I., Chen, R. C. and Chung, C. H., 2002. A target-costing based strategic decision support system. *Journal of Computer Information Systems*, 43(1), 110–116.

Liker, J., 2004. *The Toyota Way: 14 Management Principles from the World's Greatest Manufacturer*. New York: McGraw-Hill Education.

Maskell, B. H., 2012. *Practical Lean Accounting*. Boca Raton, FL: CRC Press.

Maskell, B. H. and Jenson, R., 2000. Lean accounting for lean manufacturers. *Manufacturing Engineering*, 125(6), 46–53.

Mather, H., 1986. *Competitive Manufacturing*. Upper Saddle River, NY: Prentice Hall.

Mitchell, F., Nørreklit, H. and Jakobsen, M., 2013. *The Routledge Companion to Cost Management*. New York: Taylor & Francis.

Monden, Y., 1992. *Cost Management in the New Manufacturing Age: Innovations in the Japanese Automotive Industry*. New York: Productivity Press.

Monden, Y., 2000. *Japanese Cost Management*. London, UK: World Scientific, Imperial College Press.

Monden, Y., 2012. *Toyota Production System: An Integrated Approach to Just-in-Time*. Boca Raton, FL: CRC Press.

Monden, Y., Kosuga, M., Nagasaka, Y., Hiraoka, S. and Hoshi, N., 2007. *Japanese Management Accounting Today*. Singapore: World Scientific Publishing.

Murata, K. and Katayama, H., 2009. An evaluation of factory performance utilized KPI/KAI with data envelopment analysis. *Journal of the Operations Research Society of Japan*, 52(2), 204.

Nakajima, S., 1988. *Introduction to TPM: Total Productive Maintenance*. New York: Productivity Press.

Ohno, T., 1988. *Toyota Production System: Beyond Large-Scale Production*. New York: Productivity Press.

Posteucă, A., 2011. Management branding (MB): Performance improvement through contextual managerial behavior development. *International Journal of Productivity and Performance Management*, 60(5), 529–543.

Posteucă, A., 2013. Lean Green methodology: Enterprise energy management for industrial companies. *Academy of Romanian Scientists—"Productica" Scientific Session*, 5(1), 17–30.

Posteucă, A., 2015. Manufacturing cost policy deployment by systematic and systemic improvement. PhD diss., University "Politehnica" Bucharest.

Posteucă, A., 2018. *Manufacturing Cost Policy Deployment (MCPD) Transformation: Uncovering Hidden Reserves of Profitability*. New York, USA: Taylor & Francis.

Posteucă, A. and Zapciu, M., 2013. Quick changeover: Continuous improvement and production costs reduction for plastic-molding machines. *The 7th International Working Conference*, Belgrade, Serbia, 1 (June 3–7), pp. 141–147.

Posteucă, A. and Zapciu, M., 2015b. Beyond target costing: Manufacturing cost policy deployment for new products. *Applied Mechanics and Materials*, 809, 1480–1485.

Posteucă, A. and Zapciu, M., 2015c. Continuous improvement of the effectiveness of equipment driven by the dynamics of cost reduction. *Sustainable Design and Manufacturing—KES International Conference*, Seville, Spain.

Posteucă, A. and Zapciu, M., 2015d. Process innovation: Holistic scenarios to reduce total lead time. Academy of Romanian Scientists—"Productica" scientific session.

Posteucă, A. and Zapciu, M., 2015e. *Setup time and cost reduction in conditions of low volume and overcapacity*. University "Politehnica" of Bucharest, Sci. Bull., Series D, Vol. 77.

Posteucă, A. and Sakamoto, S., 2017. *Manufacturing Cost Policy Deployment (MCPD) and Methods Design Concept (MDC): The Path to Competitiveness*. New York, USA: Taylor & Francis.

Riggs, James L. and Felix, Glenn H., 1983. *Productivity by Objectives Results-Oriented Solutions to the Productivity Puzzle*. Englewood Cliff, NJ: Prentice-Hall.

Sakamoto, S., 1992a. Design concept for methods engineering. In *Maynard Industrial Engineering Handbook*. Hodson, W. K. ed. New York: McGraw-Hill.

Sakamoto, S., 2006. Methods design concept: An effective approach to profitability. *Journal of Philippine Industrial Engineering*, 3(2), 1–11.

Sakamoto, S., 2009. *Return to Work Measurement*. Norcross, GA: Industrial Engineering, p. 24.

Sakamoto, S., 2010. *Beyond World-Class Productivity: Industrial Engineering Practice and Theory*. London UK: Springer.

Schonberger, R. J., 1986. *World-Class Manufacturing*. New York: The Free Press.

Shingo, S., 1989. *Study of' Toyota Production System from Industrial Viewpoint*. Tokio, Japan Management Association.

Shirose, K., 1999. *TPM: Total Productive Maintenance: New Implementation Program in Fabrication and Assembly Industries*. Tokio: JIPM.

Stamatis, D. H., 2010. *The OEE Primer: Understanding Overall Equipment Effectiveness, Reliability, and Maintainability*. Boca Raton, FL: CRC Press.

Suzuki, T., 1994. *TPM in Process Industries*. New York: Taylor & Francis Group.

Swedish Federation of Productivity Services, 1993. *SAM Training Program*. Stockholm, Sweden: Swedish Federation of Productivity Services.

Tanaka, T., 1994. Kaizen budgeting: Toyota's cost-control system under TQC. *Journal of Cost Management for the Manufacturing Industry*, 8(Fall), 56–62.

Tapping, D., Luyster, T. and Shuker, T., 2002. *Value Stream Management: Eight Steps to Planning, Mapping, and Sustaining Lean Improvements*. New York: Taylor & Francis Group.

Taylor, Fredrick Winslow, 1911. *The Principles of Scientific Management*. New York: Harper and Brothers.

Tennant, C. and Roberts, P., 2001. Hoshin Kanri: A tool for strategic policy deployment. *Knowledge and Process Management*, 8(4), 262–269.

Tuttle, T. C., 2016. *Growing Jobs: Transforming the Way We Approach Economic Development*. Santa Barbara, California: ABC-CLIO, LLC.

Witcher, B. J., 2003. Policy management of strategy (Hoshin Kanri). *Strategic Change*, 12(2), 83–94.

Witcher, B. J. and Chau, V. S., 2007. Balanced scorecard and Hoshin Kanri: Dynamic capabilities for managing strategic fit. *Management Decision*, 45(3), 518–538.

Womack, J. P. and Jones, D. T., 1996. *Lean Thinking: Banish Waste and Create Wealth in Your Corporation*. New York: Free Press.

Womack, J. P., Jones, D. T. and Roos, D., 1990. *The Machine That Changed the World: The Story of Lean Production—Toyota's Secret Weapon in the Global Car Wars That Is Now Revolutionizing World Industry*. New York: Free Press.

Yamamoto, Y., 2013. *Kaikaku in production toward creating unique production systems*. PhD thesis, Eskilstuna, Sweden: Department of Innovation, Design and Engineering, Mälardalen University.

Yamashina, H. and Kubo, T., 2002. Manufacturing cost deployment. *International Journal of Production Research*, 40(16), 4077–4091.

Zammori, F., Braglia, M. and Frosolini, M., 2011. Stochastic overall equipment effectiveness. *International Journal of Production Research*, 49(21), 6469–6490.

Zandin, K. B., 1980. *MOST Work Measurement System*. New York: Marcel Dekker.

Zanoni, S., Ferretti, I. and Tang, O., 2006. Cost performance and bullwhip effect in a hybrid manufacturing and remanufacturing system with different control policies. *International Journal of Production Research*, 44(18–19), 3847–3862.

Zawawi, N. H. M. and Hoque, Z., 2010. Research in management accounting innovations: An overview of its recent development. *Qualitative Research in Accounting & Management*. 7(4), 505–568.

Zengin, Y. and Ada, E., 2010. Cost management through product design: Target costing approach. *International Journal of Production Research*, 48(19), 5593–5611.

Zoysa, A. D. and Herath, S. K., 2007. Standard costing in Japanese firms: Reexamination of its significance in the new manufacturing environment. *Industrial Management & Data Systems*, 107(2), 271–283.

Index